Operational Amplifier

Operational Amplifiers

OOOOOOOOOOOOOOOOOOOOOOOOOOOOOOOOOOOO

Arpad Barna
University of Hawaii

Wiley-Interscience
a Division of John Wiley & Sons, Inc.

New York · London · Sydney · Toronto

Library of Congress Catalog Card Number: 70-150608
ISBN   0-471-05030-X

Printed in the United States of America

10 9 8 7 6 5 4 3 2 1

to Hawaii

Preface

OOOOOOOOOOOOOOOOOOOOOOOOOOOOOOOOO

The availability of mass-produced operational amplifiers at a low cost has resulted in their widespread use in many applications. Because of their performance, these devices have brought about designs with unprecedented precision, speed, reliability, and reproducibility. In order to take full advantage of this potential, thorough and precise design techniques must be applied. This text offers the reader a basic understanding of the use of operational amplifiers in linear circuits fundamental to other applications.

The book developed from a senior elective course in electronic instrumentation given at the University of Hawaii. Over 200 examples and problems expand the book's scope and illustrate realistic applications. These features, and a structure aimed at easy access to the material, make this book useful both as a text and as a reference.

After a general introduction, basic properties of ideal operational amplifiers are described. Feedback is introduced in Chapter 3. The effects of feedback and of component variations on accuracy are discussed in Chapter 4. Transient response and frequency response of operational amplifiers and feedback amplifiers are summarized in Chapters 5 and 6. Stability considerations and criteria are introduced in Chapter 7, compensation techniques are described in Chapter 8. Common mode rejection, input and output impedances, and supply voltage rejection properties are summarized in Chapter 9, input currents, offset voltage, slew rate, noise, and other limitations in Chapter 10. An Appendix provides tables summarizing the results obtained in the text and lists the properties of operational amplifiers used in the examples and problems. Answers to selected problems are also given.

Honolulu, Hawaii   ARPAD BARNA
February 1971

Contents

OOOOOOOOOOOOOOOOOOOOOOOOOOOOOOOOO

1

OOOOOOOOOOOOOOOOOOOOOOOOOOOOOOOOOO

Operational Amplifiers

This chapter is intended to introduce distinctive features of *operational amplifiers* and illustrate the theme of the book.* Detailed treatment of the subject matter begins with Chapter 2.

The term operational amplifier has attained widespread acceptance during the last decade, even though operational amplifiers have been in use for a much longer time. An operational amplifier (Fig. 1.1) is characterized by an output voltage V_{out} that is proportional to the difference of the two input voltages V_p and V_n:

$$V_{out} = A(V_p - V_n). \qquad (1.1)$$

A distinctive feature of operational amplifiers is that *amplification A* is a function of frequency with a nonzero value at dc (zero

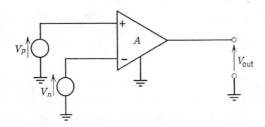

Figure 1.1.   Schematic diagram of an operational amplifier.

* New terms are introduced by italicized letters.

frequency). This value is usually the maximum of A as function of frequency.

Example 1.1. In an operational amplifier described by Equation 1.1, $A = 10,000/(1 + \text{j}f/1 \text{ MHz})$, where f is the frequency and $\text{j} \equiv \sqrt{-1}$. The dc value of A, i.e., its value at zero frequency, is $A_{DC} \equiv A(f = 0) = 10,000$. The magnitude of amplification A,

$$|A| = 10,000/\sqrt{1 + (f/1 \text{ MHz})^2}$$

has its maximum at frequency $f = 0$ and it is equal to A_{DC}.

Another distinctive feature of operational amplifiers follows from Equation 1.1: The output voltage—at least in principle—is zero when both input voltages are zero. This contrasts with other types of amplifiers, such as a single-stage transistor dc amplifier without level-shifting, where for zero input voltage the output *cannot* be zero.

PROPERTIES

Ideally, it would be desirable to have operational amplifiers described by Equation 1.1 with an amplification A that has precisely defined frequency characteristics, that is independent of signal levels, time, temperature, and is identical from unit to unit. It would be also desirable to have Equation 1.1 satisfied exactly: V_{out} should be zero when $V_p = V_n$, no matter how large they are.

In reality, amplification A is a function of signal level, time, temperature, power supply voltage, and varies from unit to unit sometimes by as much as a factor of five. Operational amplifiers also have a nonzero *input offset voltage* (typically a few millivolts) that adds to $V_p - V_n$ and that varies with temperature and from unit to unit.

Example 1.2. An operational amplifier has a dc amplification of $A_{DC} = 1000$ and a maximum input offset voltage of $V_{\text{OFF}} = \pm 2$ mV. As a result, for zero input voltages the output voltage can be anywhere within the range given by $V_{\text{out}} = V_{\text{OFF}}A = \pm 2$ mV $\times 1000 = \pm 2$ V. Thus, if $V_p = V_n = 0$, then output voltage V_{out} is between -2 V and $+2$ V.

In the case when the two input voltages are equal but not zero, that is, $V_p = V_n \neq 0$, some fraction of them will find the way to the output terminal; this fraction is designated as *common mode amplification* A_{CM}.

> **Example 1.3.** An operational amplifier has a common mode amplification of $A_{CM} = 0.0001$. Thus, if $V_p = V_n = 1$ V, the output voltage, neglecting input offset voltage V_{OFF}, will be $V_{out} = A_{CM}V_p = A_{CM}V_n = 0.0001 \times 1 \text{ V} = 0.1 \text{ mV}$.

Ideally, it would be also desirable to have operational amplifiers with zero *input currents* flowing into their two input terminals. In reality, there will always be some (possibly quite small) input current.

> **Example 1.4.** The average of the two input currents is designated as *input bias current*. An operational amplifier has a maximum input bias current of $I_B = 40 \text{ pA} = 40 \times 10^{-12}$ A. Thus, assuming that both input currents are of the same polarity, each of the two input currents could be anywhere between zero and ± 80 pA. As a result of well-controlled manufacturing technology, however, the two input currents are always within 15 pA of each other; this is expressed by stating that the maximum *input offset current* is 15 pA.

Properties of operational amplifiers will be discussed in detail in the subsequent chapters. Whether imperfections of an operational amplifier are significant, or not, depends on the requirements of the particular application.

APPLICATIONS

Operational amplifiers can be utilized in many circuits, such as amplifiers, pulse shapers, active filters, waveform generators, comparators. Several applications are illustrated in the problems at the end of this chapter; in the subsequent chapters, however, the discussion will center on utilization as amplifier—which application is perhaps basic to all other uses.* It will be seen that by the use of negative feedback, some properties of the resulting amplifier circuit may be improved at the expense of other ones, while some characteristics cannot be altered by feedback. Thus, for example, the dc amplification of the circuit and its accuracy can be changed by feedback, but the input offset voltage cannot be improved.

It will not be feasible to discuss all possible amplifier configurations. Emphasis will be placed on the analysis of simple circuits that can be used as building blocks in larger systems.

PROBLEMS

1. Determine the value of output voltage V_{out} in Fig. 1.1, if $V_p = 1$ mV, $V_n = 1.1$ mV, and $A = 10,000$.

2. Determine the value of $|A|$ at a frequency of $f = 10$ MHz, if $A = 1000/(1 + jf/10 \text{ MHz})$.

3. Determine the phase of A at a frequency of $f = 10$ MHz, if A is as given in the preceding problem.

4. An operational amplifier has a maximum input offset voltage of $V_{OFF} = \pm 3$ mV and a dc amplification of $A_{DC} = 1000$. Determine the limits of output voltage V_{out}, if the input voltages are $V_p = 5$ mV and $V_n = 6$ mV.

5. An operational amplifier has a common mode amplification of $A_{CM} = 0.0002$. Determine the absolute value of the output voltage V_{out}, if the input voltages are $V_p = V_n = 2$ V.

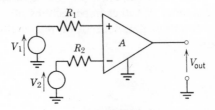

Figure 1.2.

---

* For a text on applications, and for references, see J. Eimbinder, *Application Considerations for Linear Integrated Circuits*, Wiley-Interscience, New York, 1970.

6. In the circuit of Fig. 1.2, $V_1 = 9$ mV, $V_2 = 8$ mV, $R_1 = R_2 = 10$ MΩ, and the maximum input offset current of the operational amplifier is 15 pA. Determine the limits of V_{out} if $A = 1000$ and if the amplifier obeys Equation 1.1.

7. Show that in the circuit of Fig. 1.3 ("Voltage follower"), $V_{out} = V_{in}$ if Equation 1.1 is applicable and if $A \to \infty$.

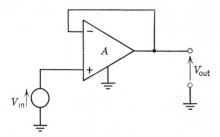

Figure 1.3.

8. An operational amplifier used as an *integrator* is shown in Fig. 1.4. Show that if the operational amplifier is described by Equation 1.1 with $A \to \infty$, and if it has zero input currents, then $V_{out} = (1/RC)\int V_{in}\, dt$ when switch S is open.

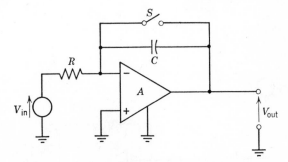

Figure 1.4.

9. Show that in the *adder* circuit of Fig. 1.5, $V_{out} = (V_1 + V_2 + V_3)/3$ if all four operational amplifiers are described by Equation 1.1 with $A \to \infty$ and if they have zero input currents.

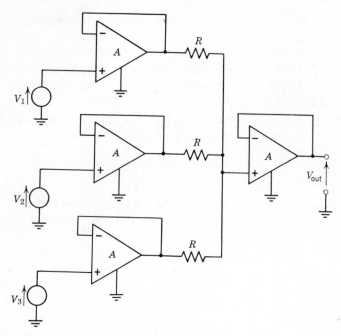

Figure 1.5.

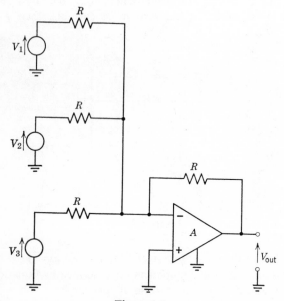

Figure 1.6.

10. Show that in the adder circuit of Fig. 1.6, $V_{out} = -(V_1 + V_2 + V_3)$, if the operational amplifier is described by Equation 1.1 with $A \to \infty$ and if it has zero input currents.

11. Modify the resistor values in the circuit of Fig. 1.6 such that $V_{out} = -(V_1 + 2V_2 + 3V_3)$. Assume that the operational amplifier is described by Equation 1.1 with $A \to \infty$ and that it has zero input currents.

12. Find V_{out} in the *pulse shaper* circuit of Fig. 1.7a, if V_{in} is as given in Fig. 1.7b. Assume that the operational amplifier is described by Equation 1.1 with $A \to \infty$ and that it has zero input currents.

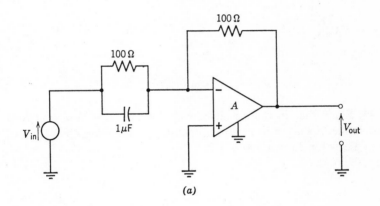

(a)

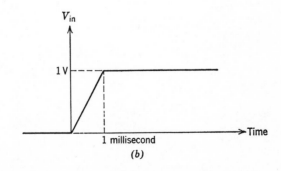

(b)

Figure 1.7.

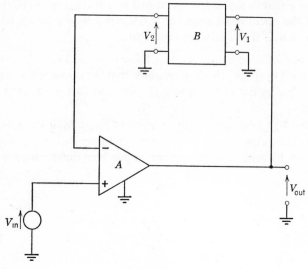

Figure 1.8.

13. Show that in the circuit of Fig. 1.8 $V_{out} = e^{V_{in}}$ if box B is described by $V_2 = \ln V_1$ and if operational amplifier A is described by Equation 1.1 with $A \rightarrow \infty$.

2

OOOOOOOOOOOOOOOOOOOOOOOOOOOOOOOOOOOO

Ideal Operational Amplifiers

An *ideal operational amplifier* (Fig. 2.1) has two input terminals, one output terminal, and one ground terminal. The voltage between the output terminal and the ground terminal, V_{out}, is related to the voltage between the positive $(+)$ terminal and ground, V_p, and to the voltage between the negative $(-)$ terminal and ground, V_n, as

$$V_{out} = A(V_p - V_n), \qquad (2.1a)$$

valid as long as the output current is finite, that is, as long as

$$|I_{out}| < \infty. \qquad (2.1b)$$

The quantity A is called *amplification, open-loop amplification,* or *differential voltage amplification* of the operational amplifier.* The

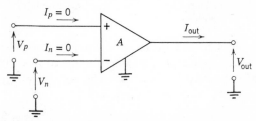

Figure 2.1. An ideal operational amplifier.

* In Chapters 2 through 4, a frequency-independent A is assumed; frequency characteristics of A are introduced in Chapter 5.

ideal operational amplifier also has zero input currents, that is, in Fig. 2.1,

$$I_p = I_n = 0. \qquad (2.2)$$

The properties described by Equations 2.1 and 2.2 are valid for any values of V_p and V_n. Thus, if, for example, an ideal operational amplifier has input voltages of $V_p = V_n = 1$ million volts, the output voltage will be $V_{out} = 0$. This follows from Equation 2.1, although in reality it would be difficult to find such an operational amplifier. It is also difficult to find Equation 2.2 satisfied in real operational amplifiers; the input currents are never zero, although in many cases they can be neglected.

NONINVERTING AMPLIFIER CIRCUITS

An ideal operational amplifier utilized as a *noninverting amplifier* is shown in Fig. 2.2. The output voltage, by using Equation 2.1, is given by

$$V_{out} = A V_{in}. \qquad (2.3)$$

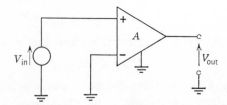

Figure 2.2.   A noninverting amplifier circuit.

Example 2.1.   An operational amplifier, used as a non-inverting amplifier, has an amplification of $A = 5000$. The input voltage is $V_{in} = 1$ mV. The output voltage is thus $V_{out} = A V_{in} = 5000 \times 1$ mV $= 5$ V.

INVERTING AMPLIFIER CIRCUITS

An ideal operational amplifier utilized as an *inverting amplifier* is shown in Fig. 2.3. The output voltage, by using Equation 2.1, is

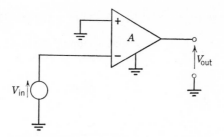

Figure 2.3. An inverting amplifier circuit.

given by

$$V_{out} = -AV_{in}. \qquad (2.4)$$

Example 2.2. An operational amplifier, used as an inverting amplifier, has an amplification of $A = 10,000$. The input voltage $V_{in} = 1$ mV. The resulting output voltage is $V_{out} = -AV_{in} = -10,000 \times 1$ mV $= -10$ V.

DIFFERENTIAL AMPLIFIER CIRCUITS

An ideal operational amplifier utilized as a *differential amplifier* is shown in Fig. 2.4. The output voltage, by using Equation 2.1, is

$$V_{out} = A(V_p - V_n). \qquad (2.5)$$

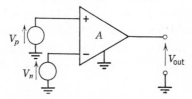

Figure 2.4. A differential amplifier circuit.

Example 2.3. An operational amplifier, used as a differential amplifier, has an amplification of $A = 20,000$. Input voltages are $V_p = 9$ mV and $V_n = 9.1$ mV. The resulting output voltage is $V_{out} = A(V_p - V_n) = 20,000 \times (9$ mV $- 9.1$ mV$) = -2$ V.

FLOATING-INPUT DIFFERENTIAL AMPLIFIER CIRCUITS

A differential amplifier with floating inputs is shown in Fig. 2.5. It follows from Equation 2.1 that the output voltage

$$V_{out} = AV_{in}. \qquad (2.6)$$

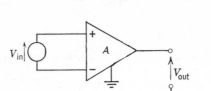

Figure 2.5. A differential amplifier circuit with floating inputs.

Example 2.4. An operational amplifier with an amplification of $A = 1000$ is used in the floating-input differential amplifier circuit of Fig. 2.5. The input voltage $V_{in} = 1$ mV. The output voltage is therefore $V_{out} = AV_{in} = 1000 \times 1$ mV $= 1$ V.

PROBLEMS

1. Determine the required value of amplification A, if an input signal with a voltage of $V_{in} = 0.1$ mV is to be amplified to a voltage of $V_{out} = 2$ V. Which amplifier circuit is to be used?

2. An inverting amplifier circuit has an amplification of $A = 5000$ and an output voltage of $V_{out} = 2$ V. Determine the value of input voltage V_{in}.

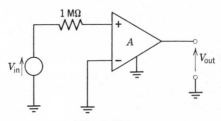

Figure 2.6.

3. Determine the value of V_p in the circuit of Fig. 2.4, if the amplification $A = 10,000$, the output voltage $V_{out} = 1$ V, and if $V_n = 5$ mV.

4. Determine the value of the output voltage V_{out} in the circuit of Fig. 2.6, if the operational amplifier is ideal with an amplification of $A = 2000$, and if $V_{in} = -1$ mV.

5. Determine the value of the output voltage V_{out} in the circuit of Fig. 2.7, if the operational amplifier is ideal with an amplification of $A = 2000$, and if $V_{in} = -1$ mV.

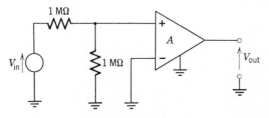

Figure 2.7.

6. Compute the values of V_{out_1} and V_{out_2} in the circuit of Fig. 2.8, assuming $A_1 = 10,000$, $A_2 = 11,000$, and $V_{in} = -0.2$ mV.

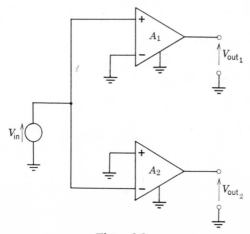

Figure 2.8.

7. Compute the value of V_{out} in the circuit of Fig. 2.9, if $V_1 = 1$ mV, $V_2 = 5$ V, $A_1 = 5000$, and $A_2 = 10{,}000$. Repeat with $A_1 = 5001$.

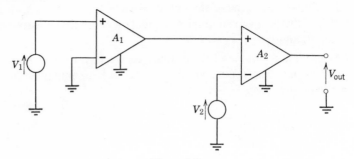

Figure 2.9.

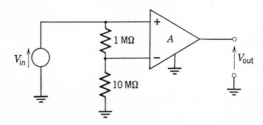

Figure 2.10.

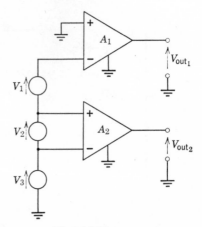

Figure 2.11.

8. Compute the value of V_{out} in the circuit of Fig. 2.10, if $V_{in} = 10$ mV and $A = 2000$.

9. Compute the values of V_{out_1} and V_{out_2} in the circuit of Fig. 2.11, if $V_1 = V_2 = V_3 = 1$ mV and $A_1 = A_2 = 4000$.

10. What is the value of V_{out} in the circuit of Fig. 2.12? Explain the result on the basis of Equation 2.1b.

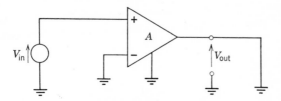

Figure 2.12.

11. The differential amplifier circuit of Fig. 2.4 is used as a comparator with $V_n = 1$ V and $V_p = t \times 1$ V/second. At what time will V_{out} equal zero?

3

OOOOOOOOOOOOOOOOOOOOOOOOOOOOOOOOOO

Feedback

The characteristics of an amplifier circuit using an operational amplifier can be substantially modified by the application of suitable *feedback:* the resulting amplification can be altered, its stability improved, the magnitude of spurious signals reduced, the bandwidth and the speed of operation increased, and nonlinearities diminished. The discussion will be focused on *negative feedback*, that is, on the case when a fraction of the output voltage is returned to the negative input terminal; some properties of *positive feedback* will be briefly mentioned in Problem 16 of this chapter and in Problem 12 of Chapter 4.

   In this chapter, amplifications of various feedback amplifier circuits will be determined; other properties of feedback will be discussed in the subsequent chapters.

NONINVERTING FEEDBACK AMPLIFIER CIRCUITS

Consider the circuit of Fig. 3.1. The input signal, V_{in}, is entered on the positive $(+)$ input terminal of the amplifier. The negative $(-)$ input terminal, however, is not connected to ground as previously, but it receives a voltage V_m that is a fraction of the output voltage V_{out}:

$$V_m = F_N V_{out}, \qquad (3.1)$$

where the *feedback return of the noninverting amplifier circuit, F_N,*

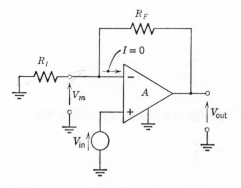

Figure 3.1. A noninverting amplifier circuit with negative feedback.

is defined as

$$F_N \equiv \frac{R_I}{R_I + R_F} . \tag{3.2}$$

In order to find V_{out} as function of V_{in}, Equation 2.1 will be applied as

$$V_{out} = A(V_{in} - V_m). \tag{3.3}$$

Combining Equations 3.1 and 3.3 results in

$$\frac{V_{out}}{V_{in}} = \frac{A}{1 + AF_N} . \tag{3.4}$$

The quantity V_{out}/V_{in} is the *resulting amplification of the noninverting feedback amplifier circuit* which in the following will be denoted by M_N,*

$$M_N \equiv \frac{V_{out}}{V_{in}} . \tag{3.5}$$

Thus, for the noninverting amplifier circuit with negative feedback,

$$M_N = \frac{A}{1 + AF_N} , \tag{3.6}$$

where the feedback return F_N is given by Equation 3.2.

* Other terms in use for M_N are *feedback amplification, closed-loop amplification.*

Example 3.1. The feedback amplifier circuit of Fig. 3.1 uses an operational amplifier with an amplification of $A = 1000$. Resistor values are $R_I = 1000 \ \Omega$ and $R_F = 9000 \ \Omega$. Thus, the feedback return

$$F_N = \frac{R_I}{R_I + R_F} = \frac{1000 \ \Omega}{1000 \ \Omega + 9000 \ \Omega} = 0.1,$$

and the resulting amplification of the feedback amplifier circuit

$$M_N = \frac{A}{1 + AF_N} = \frac{1000}{1 + 1000 \times 0.1} = \frac{1000}{101} = 9.90.$$

It is of interest to determine what happens when the amplification of the operational amplifier A is sufficiently large, so that the *feedback factor of the noninverting amplifier circuit*, defined as AF_N, becomes much larger than unity, that is, when

$$AF_N \gg 1. \tag{3.7}$$

In this case, the resulting amplification of Equation 3.6 can be simplified as

$$M_N = \frac{A}{1 + AF_N} \approx \frac{A}{AF_N} = \frac{1}{F_N}. \tag{3.8}$$

Thus, if A is sufficiently large to result in a feedback factor of $AF_N \gg 1$, the resulting amplification of the feedback amplifier circuit, M_N, becomes independent of A and is determined solely by the feedback return, that is, by resistors R_I and R_F.

Example 3.2. The feedback amplifier circuit of Fig. 3.1 uses an operational amplifier with an amplification of $A = 100,000$. Resistor values are $R_I = 1000 \ \Omega$ and $R_F = 9000 \ \Omega$. Thus, the feedback return is

$$F_N = \frac{R_I}{R_I + R_F} = \frac{1000 \ \Omega}{1000 \ \Omega + 9000 \ \Omega} = 0.1.$$

The value of the feedback factor $AF_N = 100,000 \times 0.1 = 10,000 \gg 1$. Hence, the resulting amplification of the feedback amplifier circuit is approximately $M_N \approx 1/F_N = 1/0.1 = 10$. The exact value of M_N is

$$M_N = \frac{A}{1 + AF_N} = \frac{100,000}{1 + 10,000} = 9.9990.$$

INVERTING FEEDBACK AMPLIFIER CIRCUITS

An inverting amplifier circuit with negative feedback is shown in Fig. 3.2. Here, as in the case of the noninverting amplifier circuit, a fraction of the output voltage is returned to the negative input terminal. Now, however, the input signal, V_{in}, is entered at the negative terminal via input resistor R_I. By utilizing Equation 2.1,

$$V_{out} = -AV_m. \tag{3.9}$$

Also, by inspection of Fig. 3.2,

$$I_m = \frac{V_{in} - V_m}{R_I} \tag{3.10}$$

and

$$I_m = \frac{V_m - V_{out}}{R_F}. \tag{3.11}$$

By combining Equations 3.9, 3.10, and 3.11, the *resulting amplification of the inverting feedback amplifier circuit*, M_I,* becomes

$$M_I \equiv \frac{V_{out}}{V_{in}} = \frac{-A}{1 + (A + 1)F_I}, \tag{3.12}$$

where the *feedback return of the inverting amplifier circuit*, F_I, is defined as

$$F_I \equiv \frac{R_I}{R_F}. \tag{3.13}$$

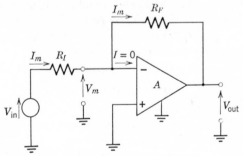

Figure 3.2. An inverting amplifier circuit with negative feedback.

* Other terms in use for M_I are *feedback amplification, closed-loop amplification.*

Example 3.3. The feedback amplifier circuit of Fig. 3.2 uses an operational amplifier with an amplification of $A = 1000$.   Resistor values are $R_I = 1000\,\Omega$ and $R_F = 10,000\,\Omega$. Thus, the feedback return

$$F_I = \frac{R_I}{R_F} = \frac{1000}{10,000} = 0.1,$$

and the resulting amplification of the feedback amplifier circuit

$$M_I = \frac{-A}{1 + (A + 1)F_I} = \frac{-1000}{1 + (1000 + 1) \times 0.1} = -9.89.$$

When amplification A is sufficiently large, so that the *feedback factor of the inverting amplifier circuit*, defined as AF_I, is large, that is, when

$$AF_I \gg 1 + F_I, \tag{3.14}$$

then Equation 3.12 can be simplified as

$$M_I = \frac{-A}{1 + (A + 1)F_I} = \frac{-A}{1 + F_I + AF_I} \approx \frac{-A}{AF_I} = -\frac{1}{F_I}. \tag{3.15}$$

Thus, as in the case of the noninverting amplifier circuit with negative feedback, if amplification A is sufficiently large, the resulting amplification is determined solely by the feedback return, that is, by resistors R_I and R_F.

Example 3.4. The feedback amplifier circuit of Fig. 3.2 uses an operational amplifier with an amplification of $A = 100,000$.   Resistor values are $R_I = 1000\,\Omega$ and $R_F = 10,000\,\Omega$. Thus, the feedback return is

$$F_I = \frac{R_I}{R_F} = \frac{1000}{10,000} = 0.1.$$

The value of the feedback factor $AF_I = 100,000 \times 0.1 = 10,000 \gg 1 + F_I = 1.1$.   Therefore, the resulting amplification of the feedback amplifier circuit is approximately

$M_I \approx -1/F_I = -1/0.1 = -10.$ The exact value of M_I is

$$M_I = \frac{-A}{1 + (A + 1)F_I} = \frac{-100,000}{1 + (100,000 + 1) \times 0.1}$$

$$= -9.9989.$$

DIFFERENTIAL FEEDBACK AMPLIFIER CIRCUITS

A differential amplifier circuit with negative feedback is shown in Fig. 3.3. The following equations can be written:

$$V_{\text{out}} = A(V_p - V_m), \tag{3.16}$$

$$I_m = \frac{V_n - V_m}{R_I}, \tag{3.17}$$

and

$$I_m = \frac{V_m - V_{\text{out}}}{R_F}. \tag{3.18}$$

The combination of these equations results in

$$V_{\text{out}} = \frac{A}{1 + AF_N} V_p - \frac{A}{1 + (A + 1)F_I} V_n, \tag{3.19}$$

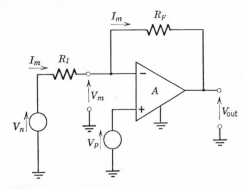

Figure 3.3 A differential amplifier circuit with negative feedback.

where feedback returns F_N and F_I are defined as

$$F_N \equiv \frac{R_I}{R_I + R_F} \qquad (3.20)$$

and

$$F_I \equiv \frac{R_I}{R_F}. \qquad (3.21)$$

Equation 3.19 can be also written as

$$V_{out} = M_N V_p + M_I V_n, \qquad (3.22)$$

where M_N and M_I are defined as

$$M_N \equiv \frac{A}{1 + AF_N} \qquad (3.23)$$

and

$$M_I \equiv \frac{-A}{1 + (A + 1)F_I} \qquad (3.24)$$

with F_N and F_I given by Equations 3.20 and 3.21, respectively.

It can be seen that, in general, the magnitudes of M_N and M_I are not equal; therefore, the two input signals V_p and V_n are amplified by different factors. In the limiting case when feedback factors $AF_N \gg 1$ and $AF_I \gg 1 + F_I$,* output voltage V_{out} can be approximated as

$$V_{out} \approx \frac{1}{F_N} V_p - \frac{1}{F_I} V_n. \qquad (3.25)$$

Example 3.5. The circuit of Fig. 3.3 uses an operational amplifier with an amplification of $A = 100,000$. Resistor values are $R_I = 1000 \ \Omega$ and $R_F = 9000 \ \Omega$. The value of feedback return F_N is

$$F_N = \frac{R_I}{R_I + R_F} = \frac{1000 \ \Omega}{1000 \ \Omega + 9000 \ \Omega} = \frac{1}{10}$$

and the value of feedback return F_I is

$$F_I = \frac{R_I}{R_F} = \frac{1000 \ \Omega}{9000 \ \Omega} = \frac{1}{9}.$$

* It can be shown that these two conditions are equivalent.

The value of feedback factor AF_N is

$$AF_N = 100{,}000/10 = 10{,}000 \gg 1,$$

and the use of Equation 3.25 is justified:

$$V_{\text{out}} \approx \frac{1}{F_N} V_p - \frac{1}{F_I} V_n = 10V_p - 9V_n.$$

The exact expression for V_{out}, by utilizing Equation 3.19, is

$$V_{\text{out}} = \frac{A}{1 + AF_N} V_p - \frac{A}{1 + (A+1)F_I} V_n$$

$$= \frac{100{,}000}{1 + 100{,}000 \times 0.1} V_p - \frac{100{,}000}{1 + (100{,}000 + 1)/9} V_n$$

$$= 9.9990 V_p - 8.9991 V_n.$$

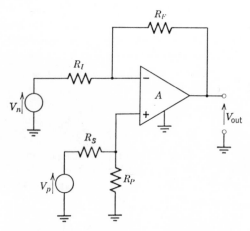

Figure 3.4. A differential amplifier circuit with negative feedback and equalized amplifications.

The magnitudes of the amplifications of the two input signals V_p and V_n in the differential feedback amplifier circuit can be made equal by modifying the circuit of Fig. 3.3 as shown in Fig. 3.4. By utilizing

Equation 3.19, the output voltage V_{out} can be now written as

$$V_{\text{out}} = \frac{A}{1 + AF_N} \frac{R_P}{R_P + R_S} V_p - \frac{A}{1 + (A + 1)F_I} V_n$$

$$= \frac{A}{1 + \dfrac{R_S}{R_P} + A\dfrac{R_I}{R_F}\left(1 + \dfrac{R_S}{R_P}\right) \Big/ \left(1 + \dfrac{R_I}{R_F}\right)} V_p$$

$$- \frac{A}{1 + \dfrac{R_I}{R_F} + A\dfrac{R_I}{R_F}} V_n. \tag{3.26}$$

If

$$\frac{R_P}{R_S} = \frac{R_F}{R_I}, \tag{3.27}$$

then Equation 3.26 becomes

$$V_{\text{out}} = \frac{A}{1 + (A + 1)F_I} (V_p - V_n), \tag{3.28}$$

that is, the magnitudes of the amplifications of V_n and V_p are equal. If, furthermore, feedback factors $AF_N \gg 1$ and $AF_I \gg 1 + F_I$, then V_{out} can be approximated as

$$V_{\text{out}} \approx \frac{1}{F_I} (V_p - V_n). \tag{3.29}$$

Example 3.6. In the circuit of Fig. 3.4, the amplification of the operational amplifier is $A = 100{,}000$. Resistor values are $R_I = 1000\ \Omega$, $R_F = 9000\ \Omega$, $R_S = 2000\ \Omega$, and $R_P = 18{,}000\ \Omega$. Hence, $R_P/R_S = R_F/R_I$, and feedback factor

$$AF_I = A\frac{R_I}{R_F} = 100{,}000\frac{1000\ \Omega}{9000\ \Omega} = 11{,}111 \gg 1 + F_I$$

$$= 1 + \frac{R_I}{R_F} = 1.11.$$

Thus, Equation 3.29 is applicable, i.e., the output voltage is approximately

$$V_{\text{out}} \approx \frac{1}{F_I} (V_p - V_n) = 9(V_p - V_n).$$

The exact expression for V_{out}, by utilizing Equation 3.28, is

$$V_{\text{out}} = \frac{A}{1 + (A + 1)F_I} (V_p - V_n)$$

$$= \frac{100,000}{1 + (100,000 + 1)/9}(V_p - V_n) = 8.9991(V_p - V_n).$$

VOLTAGE FOLLOWER CIRCUITS

A special case of the noninverting feedback amplifier circuit of Fig. 3.1 occurs in the limit when $R_I \neq 0$ and $R_F = 0$, or when $R_I = \infty$ and $R_F \neq \infty$. This special case is the *voltage follower circuit*, the simplest form of which is shown in Fig. 3.5. The value of feedback return $F_N = 1$; thus, by utilizing Equation 3.4,

$$\frac{V_{\text{out}}}{V_{\text{in}}} = \frac{A}{1 + A}. \tag{3.30}$$

In the limiting case when amplification $A \gg 1$, Equation 3.30 reduces to

$$\frac{V_{\text{out}}}{V_{\text{in}}} \approx 1, \tag{3.31}$$

whence the name voltage follower.

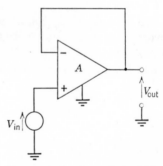

Figure 3.5. A voltage follower circuit.

Example 3.7. The circuit of Fig. 3.5 uses an operational amplifier with an amplification of $A = 100,000$. Therefore, by utilizing Equation 3.30,

$$\frac{V_{\text{out}}}{V_{\text{in}}} = \frac{100,000}{1 + 100,000} = 0.999990.$$

PROBLEMS

1. Compute the value of the resulting amplification of a noninverting feedback amplifier circuit, M_N, if the operational amplifier used in the circuit has an amplification of $A = 10,000$; $R_I = 100 \, \Omega$, and $R_F = 10,000 \, \Omega$. What is the fractional change of M_N in percents, if A is changed to 11,000?

2. Derive V_m/V_{in} in Fig. 3.1. What is V_m/V_{in} if feedback factor $AF_N \gg 1$?

3. Derive Equation 3.4.

4. Compute the value of the resulting amplification of an inverting feedback amplifier circuit, M_I, in the circuit of Fig. 3.2, if the operational amplifier has an amplification of $A = 10,000$; $R_I = 100 \, \Omega$, and $R_F = 10,000 \, \Omega$.

5. Derive V_m/V_{in} in Fig. 3.2. What is V_m/V_{in} if feedback factor $AF_I \gg 1 + F_I$?

6. Derive Equation 3.12.

7. Derive V_{in}/I_m in Fig. 3.2. What is V_{in}/I_m if feedback factor $AF_I \gg 1 + F_I$?

8. Compute the values of the resulting amplifications M_N and M_I of a differential amplifier circuit with feedback (Fig. 3.3), if the operational amplifier used in the circuit has an amplification of $A = 10,000$; $R_I = 100 \, \Omega$, and $R_F = 10,000 \, \Omega$.

9. Derive Equation 3.19.

10. In the circuit of Fig. 3.4, the operational amplifier has an amplification of $A = 10,000$. Resistor values are $R_I = 100 \, \Omega$, $R_F = 10,000 \, \Omega$, $R_S = 100 \, \Omega$, and $R_P = 10,000 \, \Omega$; input voltages are $V_p = V_n = 1$ V. What is the value of output voltage V_{out}? Repeat with $R_P = 10,001 \, \Omega$.

11. Derive Equation 3.28.

12. Compute the value of $V_{\text{out}}/V_{\text{in}}$ in a voltage follower circuit using an operational amplifier with an amplification of $A = 1000$.

13. Compute V_{out}/V_{in} in the circuit of Fig. 3.6. Comment on the result.

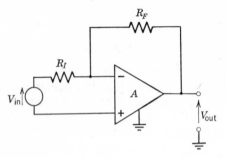

Figure 3.6.

14. Compute the value of output voltage V_{out} in the circuit of Fig. 3.7, if $V_1 = 1$ mV and amplification $A = 1000$.

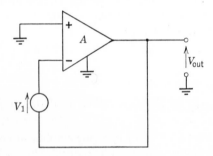

Figure 3.7.

15. Derive an expression for V_{out} in the *compound differential amplifier* circuit of Fig. 3.8. How does V_{out} depend on R_1 and R_2? Evaluate V_{out} for $V_1 = 10$ mV, $V_2 = 11$ mV, $R_1 = 1$ MΩ, $R_2 = 10$ MΩ, $A_1 = 11,000$, $A_2 = 10,000$, $R_I = R_S = 100 \Omega$, $R_F = R_P = 100,000 \Omega$, and $A_3 = 10,000$. What is the purpose of operational amplifiers A_1 and A_2?

16. An amplifier circuit with *positive feedback* is shown in Fig. 3.9. Determine the value of V_{out}/V_{in}, if the amplification of the operational amplifier $A = 100$; $R_I = 100 \Omega$, and $R_F = 10,000 \Omega$. Comment on the result.

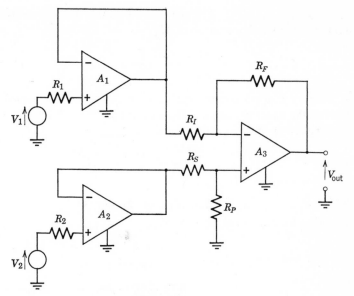

Figure 3.8.

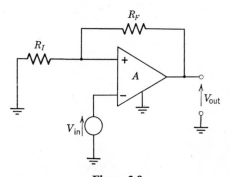

Figure 3.9.

17. When the approximate expression of $M_N \approx 1/F_N$ is used in place of the exact Equation 3.6, an error in M_N will result. The fractional error of M_N is defined as

$$\frac{\Delta M_N}{M_N} \equiv \frac{M_{N_{\text{exact}}} - M_{N_{\text{approx}}}}{M_{N_{\text{exact}}}},$$

where M_{N_exact} is M_N of Equation 3.6, and $M_{N_\text{approx}} = 1/F_N$. Show that the fractional error of M_N,

$$\frac{\Delta M_N}{M_N} = - \frac{1}{AF_N}.$$

18. When the approximate expression of $M_I \approx -1/F_I$ is used in place of the exact Equation 3.12, an error in M_I will result. The fractional error of M_I is defined as

$$\frac{\Delta M_I}{M_I} \equiv \frac{M_{I_\text{exact}} - M_{I_\text{approx}}}{M_{I_\text{exact}}},$$

where M_{I_exact} is M_I of Equation (3.12) and $M_{I_\text{approx}} = -1/F_I$. Show that the fractional error of M_I,

$$\frac{\Delta M_I}{M_I} = - \frac{1 + F_I}{AF_I}.$$

19. Determine V_out/V_in in the *differential-in differential-out amplifier* circuit of Fig. 3.10.* Evaluate V_out/V_in for $A_1 = A_2 = 1000$, $R_1 = R_2 = 10,000\ \Omega$, and $R_3 = 2000\ \Omega$.

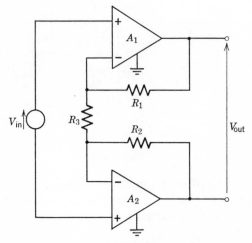

Figure 3.10.

See footnote on p. 31.

20. Determine V_{out}/V_{in} in the *potentiometric amplifier* circuit of Fig. 3.11.* Evaluate V_{out}/V_{in} for $A_1 = A_2 = 1000$ and $R_1 = R_2 = R_3 = R_4 = 1000 \ \Omega$.

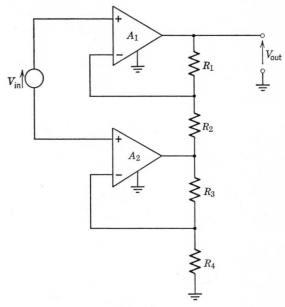

Figure 3.11.

21. Show that conditions $AF_N \gg 1$ and $AF_I \gg 1 + F_I$ are equivalent, if F_N and F_I are given by Equations 3.20 and 3.21, respectively.

* For a detailed analysis of the circuit see J. Eimbinder, *Designing with Linear Integrated Circuits*, John Wiley and Sons, New York, 1969.

4

OOOOOOOOOOOOOOOOOOOOOOOOOOOOOOOOOO

Accuracy of the Amplification

Expressions for the resulting amplifications M_N and M_I derived in the preceding chapter show that they are functions of the amplification of the operational amplifier A and of resistor values R_I and R_F. It is frequently necessary to evaluate the effects on M_N and M_I of small changes in amplification A and in resistor values. This could be, of course, always performed by evaluating M_N and M_I; this process, however, can become quite tedious, particularly for small changes.

Example 4.1. A noninverting feedback amplifier circuit with negative feedback (Fig 3.1) uses an operational amplifier with an amplification of $A = 10,000 \pm 1\%$. Resistor values are $R_I = 1000\,\Omega$ and $R_F = 9000\,\Omega$. What is the fractional change in the resulting amplification of the feedback amplifier circuit M_N, as a result of the 1% change in A?

The value of feedback return F_N is

$$F_N = \frac{R_I}{R_I + R_F} = \frac{1000\,\Omega}{1000\,\Omega + 9000\,\Omega} = 0.1.$$

At the nominal value of $A = 10,000$, M_N becomes

$$M_{N_{\text{nom}}} = \frac{A}{1 + AF_N} = \frac{10,000}{1 + 10,000 \times 0.1} = \frac{10,000}{1001} \approx 9.99.$$

33

At the minimum value of $A = 9900$, M_N is

$$M_{N_{\min}} = \frac{A}{1 + AF_N} = \frac{9900}{1 + 9900 \times 0.1} = \frac{9900}{991}.$$

At the maximum value of $A = 10,100$, M_N becomes

$$M_{N_{\max}} = \frac{A}{1 + AF_N} = \frac{10,100}{1 + 10,100 \times 0.1} = \frac{10,100}{1011}.$$

The difference between the minimum and the nominal values of M_N is

$$M_{N_{\min}} - M_{N_{\text{nom}}} = \frac{9900}{991} - \frac{10,000}{1001} = -\frac{100}{991,991} \approx -0.0001$$

and the fractional change in M_N as a result of this is

$$\frac{M_{N_{\min}} - M_{N_{\text{nom}}}}{M_{N_{\text{nom}}}} \approx \frac{-0.0001}{9.99} \approx -0.00001 = -0.001\,\%.$$

The difference between the maximum and the nominal values of M_N is

$$M_{N_{\max}} - M_{N_{\text{nom}}} \approx \frac{10,100}{1011} - \frac{10,000}{1001} = \frac{100}{1,012,011} \approx -0.0001$$

and the fractional change in M_N as a result of this is

$$\frac{M_{N_{\max}} - M_{N_{\text{nom}}}}{M_{N_{\text{nom}}}} \approx \frac{0.0001}{9.99} \approx 0.00001 = 0.001\,\%.$$

Thus, the resulting amplification can be written as

$$M_N = 9.99 \pm 0.001\,\%.$$

SMALL VARIATIONS IN OPERATIONAL AMPLIFIER AMPLIFICATION

The preceding example illustrates how cumbersome the evaluation of the change in the resulting amplification M_N can become if variations in amplification A of the operational amplifier are small. It will be shown now that, in the case when the magnitudes of the variations in A are much smaller than A, simple expressions can be obtained for the fractional changes in M_N and M_I.*

By defining

$$\Delta A \equiv A - A_{\mathrm{nom}} \tag{4.1}$$

and

$$\Delta M \equiv M - M_{\mathrm{nom}}, \tag{4.2}$$

where

$$M_{\mathrm{nom}} \equiv M(A_{\mathrm{nom}}), \tag{4.3}$$

and by assuming that $\Delta A / A$ is small, that is,

$$\left| \frac{\Delta A}{A} \right| \ll 1, \tag{4.4}$$

the resulting amplification M can be expanded in a Taylor series as

$$M = M_{\mathrm{nom}} + \left(\frac{\partial M}{\partial A} \right)_{A=A_{\mathrm{nom}}} (A - A_{\mathrm{nom}})$$
$$+ \frac{1}{2!} \left(\frac{\partial^2 M}{\partial A^2} \right)_{A=A_{\mathrm{nom}}} (A - A_{\mathrm{nom}})^2 + \cdots. \tag{4.5}$$

The combination of Equations 4.1, 4.2, 4.3, and 4.5 results in

$$\Delta M = \left(\frac{\partial M}{\partial A} \right)_{A=A_{\mathrm{nom}}} \Delta A + \frac{1}{2!} \left(\frac{\partial^2 M}{\partial A^2} \right)_{A=A_{\mathrm{nom}}} \Delta A^2 + \cdots. \tag{4.6}$$

It can be shown that, as a result of Equation 4.4, Equation 4.6 can be approximated as

$$\Delta M = \left(\frac{\partial M}{\partial A} \right)_{A=A_{\mathrm{nom}}} \Delta A, \tag{4.7}$$

and thus, the fractional change in the resulting amplification

$$\frac{\Delta M}{M_{\mathrm{nom}}} = \frac{A_{\mathrm{nom}}}{M_{\mathrm{nom}}} \left(\frac{\partial M}{\partial A} \right)_{A=A_{\mathrm{nom}}} \frac{\Delta A}{A_{\mathrm{nom}}}. \tag{4.8}$$

In the case of a noninverting feedback amplifier circuit, by applying Equation 4.8 to Equation 3.6 with $M = M_N$,

$$\frac{\Delta M_N}{M_{N_{\mathrm{nom}}}} = \frac{1}{1 + A_{\mathrm{nom}} F_N} \frac{\Delta A}{A_{\mathrm{nom}}} \tag{4.9}$$

* The results in this chapter will be derived by using Taylor series expansions. It is also possible, however, to arrive at these directly—see Problems 13 and 14.

results. When feedback factor $A_{nom}F_N \gg 1$, Equation 4.9 reduces to

$$\frac{\Delta M_N}{M_{N_{nom}}} \approx \frac{1}{A_{nom}F_N} \frac{\Delta A}{A_{nom}}. \tag{4.10}$$

Example 4.2. The expression of Equation 4.10 will be applied to calculate again the fractional change of M_N for the preceding example, i.e., for $A = 10{,}000 \pm 1\%$ and $F_N = 0.1$. The value of feedback factor $A_{nom}F_N = 10{,}000 \times 0.1 = 1000 \gg 1$; hence, the use of Equation 4.10 is justified:

$$\frac{\Delta M_N}{M_{N_{nom}}} = \frac{1}{A_{nom}F_N} \frac{\Delta A}{A_{nom}} = \frac{1}{1000} 1\% = 0.001\%.$$

The nominal value of the resulting amplification of the feedback amplifier circuit

$$M_{N_{nom}} = \frac{A_{nom}}{1 + A_{nom}F_N} \approx 9.99,$$

thus, M_N can be written as

$$M_N = 9.99 \pm 0.001\%.$$

In the case of an inverting feedback amplifier circuit (Fig. 3.2), by applying Equation 4.8 to Equation 3.12 with $M = M_I$,

$$\frac{\Delta M_I}{M_{I_{nom}}} = \frac{1 + F_I}{1 + (A_{nom} + 1)F_I} \frac{\Delta A}{A_{nom}} \tag{4.11}$$

results. When feedback factor $A_{nom}F_I \gg 1 + F_I$, Equation 4.11 reduces to

$$\frac{\Delta M_I}{M_{I_{nom}}} \approx \frac{1 + F_I}{A_{nom}F_I} \frac{\Delta A}{A_{nom}}. \tag{4.12}$$

Example 4.3. An inverting amplifier circuit with negative feedback uses an operational amplifier with an amplification of $A = 10{,}000 \pm 1\%$. Resistor values are $R_I = 1000\ \Omega$ and $R_F = 10{,}000\ \Omega$. What is the nominal value of M_I and its fractional change as a result of the 1% error in A?

From Equation 3.13,

$$F_I = \frac{R_I}{R_F} = \frac{1000 \ \Omega}{10,000 \ \Omega} = 0.1.$$

By applying Equation 3.12, the nominal value of M_I is

$$M_{I_{\text{nom}}} = \frac{-A_{\text{nom}}}{1 + (A_{\text{nom}} + 1)F_I}$$

$$= \frac{-10,000}{1 + (10,000 + 1) \times 0.1} \approx -9.99.$$

The value of feedback factor $A_{\text{nom}}F_I = 10,000 \times 0.1 = 1000 \gg 1 + F_I = 1.1$; hence, Equation 4.12 is applicable:

$$\frac{\Delta M_I}{M_{I_{\text{nom}}}} = \frac{1 + F_I}{A_{\text{nom}}F_I} \frac{\Delta A}{A_{\text{nom}}} = \frac{1 + 0.1}{1000} 1\% = 0.0011\%.$$

Thus, $M_I = -9.99 \pm 0.0011\%$.

SMALL VARIATIONS IN THE FEEDBACK RESISTOR

Assume next that, in Fig. 3.1 or in Fig. 3.2, amplification A of the operational amplifier and input resistor R_I are constant and only feedback resistor R_F varies. In this case,

$$M = M(R_{F_{\text{nom}}}) + \left(\frac{\partial M}{\partial R_F}\right)_{R_F = R_{F_{\text{nom}}}} (R_F - R_{F_{\text{nom}}})$$

$$+ \frac{1}{2!} \left(\frac{\partial^2 M}{\partial R_F^2}\right)_{R_F = R_{F_{\text{nom}}}} (R_F - R_{F_{\text{nom}}})^2 + \cdots. \quad (4.13)$$

By defining

$$\Delta R_F \equiv R_F - R_{F_{\text{nom}}} \quad (4.14)$$

and utilizing the definition given by Equation 4.2 with

$$M_{\text{nom}} \equiv M(R_{F_{\text{nom}}}), \quad (4.15)$$

Equation 4.13 becomes

$$\Delta M = \left(\frac{\partial M}{\partial R_F}\right)_{R_F = R_{F_{\text{nom}}}} \Delta R_F$$

$$+ \frac{1}{2!} \left(\frac{\partial^2 M}{\partial R_F^2}\right)_{R_F = R_{F_{\text{nom}}}} \Delta R_F^2 + \cdots. \quad (4.16)$$

It can be shown that for small fractional variations in the feedback resistor R_F, that is, for

$$\left| \frac{\Delta R_F}{R_{F_{\text{nom}}}} \right| \ll 1, \tag{4.17}$$

Equation 4.16 can be approximated as

$$\Delta M = \left(\frac{\partial M}{\partial R_F} \right)_{R_F = R_{F_{\text{nom}}}} \Delta R_F, \tag{4.18}$$

and thus, the fractional change in the resulting amplification

$$\frac{\Delta M}{M_{\text{nom}}} = \frac{R_{F_{\text{nom}}}}{M_{\text{nom}}} \left(\frac{\partial M}{\partial R_F} \right)_{R_F = R_{F_{\text{nom}}}} \frac{\Delta R_F}{R_{F_{\text{nom}}}}. \tag{4.19}$$

In the case of a noninverting feedback amplifier circuit, by applying Equations 3.2 and 3.6, to Equation 4.19 with $M = M_N$,

$$\frac{\Delta M_N}{M_{N_{\text{nom}}}} = M_{N_{\text{nom}}} F_{N_{\text{nom}}} (1 - F_{N_{\text{nom}}}) \frac{\Delta R_F}{R_{F_{\text{nom}}}}, \tag{4.20}$$

results, where feedback return $F_{N_{\text{nom}}}$ is defined as

$$F_{N_{\text{nom}}} \equiv (F_N)_{R_F = R_{F_{\text{nom}}}} = \frac{R_I}{R_I + R_{F_{\text{nom}}}}. \tag{4.21}$$

In the limiting case when feedback factor $AF_{N_{\text{nom}}} \gg 1$, $M_{N_{\text{nom}}} \approx 1/F_{N_{\text{nom}}}$ (see Equation 3.8); hence, the fractional change in the resulting amplification becomes

$$\frac{\Delta M_N}{M_{N_{\text{nom}}}} \approx (1 - F_{N_{\text{nom}}}) \frac{\Delta R_F}{R_{F_{\text{nom}}}}. \tag{4.22}$$

Example 4.4. A noninverting feedback amplifier circuit with negative feedback uses an operational amplifier with an amplification of $A = 10,000$. Resistor values are $R_I = 1000 \, \Omega$ and $R_F = 9000 \, \Omega \pm 1\%$. What is the fractional change in the resulting amplification M_N, as a result of the 1% change in feedback resistor R_F?

The nominal value of feedback return F_N is

$$F_{N_{\text{nom}}} = \frac{R_I}{R_I + R_{F_{\text{nom}}}} = \frac{1000 \, \Omega}{1000 \, \Omega + 9000 \, \Omega} = 0.1.$$

The value of feedback factor $AF_{N_{\text{nom}}} = 10,000 \times 0.1 = 1000 \gg 1$; hence, Equation 4.22 is applicable:

$$\frac{\Delta M_N}{M_{N_{\text{nom}}}} \approx (1 - F_{N_{\text{nom}}}) \frac{\Delta R_F}{R_{F_{\text{nom}}}} = (1 - 0.1) \times 1\% = 0.9\%.$$

In the case of an inverting amplifier circuit with $A \gg 1$, it can be shown, by utilizing Equations 3.12 and 3.13, that with $M = M_I$ Equation 4.19 becomes

$$\frac{\Delta M_I}{M_{I_{\text{nom}}}} = -M_{I_{\text{nom}}} F_{I_{\text{nom}}} \frac{\Delta R_F}{R_{F_{\text{nom}}}}, \qquad (4.23)$$

where feedback return $F_{I_{\text{nom}}}$ is defined as

$$F_{I_{\text{nom}}} \equiv (F_I)_{R_F = R_{F_{\text{nom}}}} = \frac{R_I}{R_{F_{\text{nom}}}}. \qquad (4.24)$$

In the limiting case when feedback factor $AF_{I_{\text{nom}}} \gg 1 + F_{I_{\text{nom}}}$, Equation 4.23 reduces to

$$\frac{\Delta M_I}{M_{I_{\text{nom}}}} \approx \frac{\Delta R_F}{R_{F_{\text{nom}}}}. \qquad (4.25)$$

Example 4.5. An inverting feedback amplifier circuit with negative feedback uses an operational amplifier with an amplification of $A = 10,000$. Resistor values are $R_I = 1000\ \Omega$ and $R_F = 10,000\ \Omega \pm 1\%$. What is the fractional change in the resulting amplification M_N, as a result of the 1% change in feedback resistor R_F?

The nominal value of the feedback return F_I is

$$F_{I_{\text{nom}}} = \frac{R_I}{R_{F_{\text{nom}}}} = \frac{1000\ \Omega}{10,000\ \Omega} = 0.1.$$

The value of feedback factor $AF_{I_{\text{nom}}} = 10,000 \times 0.1 = 1000 \gg 1 + F_I = 1.1$; hence, Equation 4.25 is applicable:

$$\frac{\Delta M_I}{M_{I_{\text{nom}}}} \approx \frac{\Delta R_F}{R_{F_{\text{nom}}}} = 1\%.$$

Since

$$M_{I_{\text{nom}}} = \frac{-A}{1 + (A + 1)F_{I_{\text{nom}}}}$$

$$= \frac{-10{,}000}{1 + (10{,}000 + 1)0.1} \approx -9.99,$$

the resulting amplification of the feedback amplifier circuit can be written as $M_I = -9.99 \pm 1\%$.

SMALL VARIATIONS IN THE INPUT RESISTOR

In the case when the amplification A of the operational amplifier and resistor R_F are constant and only input resistor R_I varies, it can be shown that for a noninverting amplifier circuit

$$\frac{\Delta M_N}{M_{N_{\text{nom}}}} = -M_{N_{\text{nom}}} F_{N_{\text{nom}}} (1 - F_{N_{\text{nom}}}) \frac{\Delta R_I}{R_{I_{\text{nom}}}}, \qquad (4.26)$$

where feedback return $F_{N_{\text{nom}}}$ is defined by Equation 4.21, and ΔR_I is defined as

$$\Delta R_I \equiv R_I - R_{I_{\text{nom}}}. \qquad (4.27)$$

In the limit when feedback factor $AF_{N_{\text{nom}}} \gg 1$, Equation 4.26 reduces to

$$\frac{\Delta M_N}{M_{N_{\text{nom}}}} \approx -(1 - F_{N_{\text{nom}}}) \frac{\Delta R_I}{R_{I_{\text{nom}}}}. \qquad (4.28)$$

For an inverting amplifier circuit with $A \gg 1$, it can be shown that

$$\frac{\Delta M_I}{M_{I_{\text{nom}}}} = M_{I_{\text{nom}}} F_{I_{\text{nom}}} \frac{\Delta R_I}{R_{I_{\text{nom}}}}. \qquad (4.29)$$

In the limiting case when feedback factor $AF_{I_{\text{nom}}} \gg 1 + F_{I_{\text{nom}}}$, Equation 4.29 reduces to

$$\frac{\Delta M_I}{M_{I_{\text{nom}}}} \approx -\frac{\Delta R_I}{R_{I_{\text{nom}}}}. \qquad (4.30)$$

SEVERAL SOURCES OF VARIATIONS

The Taylor series expansion applied above to errors resulting from variations in A, R_F, and R_I, can be also applied to errors resulting from other sources. It can be also applied to find the error resulting from simultaneous variations in several parameters.

If the amplification A of the operational amplifier, feedback resistor R_F, and input resistor R_I vary, then the resulting amplification M of the feedback amplifier circuit can be expressed in a multi-variable Taylor series as

$$M = M_{\text{nom}} + \left(\frac{\partial M}{\partial A}\right)(A - A_{\text{nom}}) + \left(\frac{\partial M}{\partial R_F}\right)(R_F - R_{F_{\text{nom}}})$$

$$+ \left(\frac{\partial M}{\partial R_I}\right)(R_I - R_{I_{\text{nom}}}) + \cdots, \quad (4.31)$$

where all partial derivatives have to be evaluated at $A = A_{\text{nom}}$, $R_F = R_{F_{\text{nom}}}$, $R_I = R_{I_{\text{nom}}}$.

If fractional changes in A, R_F, and R_I are small, then from Equation 4.31 and with the definitions of Equations 4.1, 4.2, 4.14, and 4.27, the change in M can be approximated as

$$\Delta M \approx \left(\frac{\partial M}{\partial A}\right)\Delta A + \left(\frac{\partial M}{\partial R_F}\right)\Delta R_F + \left(\frac{\partial M}{\partial R_I}\right)\Delta R_I, \quad (4.32)$$

where the partial derivatives again have to be evaluated at $A = A_{\text{nom}}$, $R_F = R_{F_{\text{nom}}}$, $R_I = R_{I_{\text{nom}}}$.

In the case of a noninverting amplifier circuit with a feedback factor of $A_{\text{nom}}F_{N_{\text{nom}}} \gg 1$, the application of Equations 4.10, 4.22, and 4.28 to Equation 4.32 with $M = M_N$ results in a fractional change of

$$\frac{\Delta M_N}{M_N} \approx \frac{1}{A_{\text{nom}}F_{N_{\text{nom}}}}\frac{\Delta A}{A_{\text{nom}}}$$

$$+ (1 - F_{N_{\text{nom}}})\frac{\Delta R_F}{R_{F_{\text{nom}}}} - (1 - F_{N_{\text{nom}}})\frac{\Delta R_I}{R_{I_{\text{nom}}}}. \quad (4.33)$$

In many cases it is important to find the worst case $\Delta M_N / M_N$, which is the maximum of its absolute value:

$$\text{Max} \left| \frac{\Delta M_N}{M_N} \right| \approx \left| \frac{1}{A_{\text{nom}} F_{N_{\text{nom}}}} \frac{\Delta A}{A_{\text{nom}}} \right|$$

$$+ \left| (1 - F_{N_{\text{nom}}}) \frac{\Delta R_F}{R_{F_{\text{nom}}}} \right| + \left| (1 - F_{N_{\text{nom}}}) \frac{\Delta R_I}{R_{I_{\text{nom}}}} \right|. \quad (4.34)$$

Example 4.6.   A noninverting amplifier circuit with negative feedback uses an operational amplifier with an amplification of $A = 100,000 \pm 10\%$. Resistor values are $R_I = 100\,\Omega \pm 0.1\%$ and $R_F = 100,000\,\Omega \pm 0.1\%$. Hence,

$$F_{N_{\text{nom}}} = \frac{R_{I_{\text{nom}}}}{R_{I_{\text{nom}}} + R_{F_{\text{nom}}}} = \frac{100\,\Omega}{100\,\Omega + 100,000\,\Omega} \approx 0.001,$$

and feedback factor $A_{\text{nom}} F_{N_{\text{nom}}} = 100,000 \times 0.001 = 100 \gg 1$, thus, Equation 4.34 is applicable. The worst case fractional error in M_N is, therefore,

$$\text{Max} \left| \frac{\Delta M_N}{M_N} \right| \approx \left| \frac{1}{100} 10\% \right| + |(1 - 0.001) \times 0.1\%|$$

$$+ |(1 - 0.001) \times 0.1\%| \approx 0.3\%.$$

Similarly, in the case of an inverting amplifier circuit with a feedback factor of $A_{\text{nom}} F_{I_{\text{nom}}} \gg 1 + F_{I_{\text{nom}}}$, the application of Equations 4.12, 4.25, and 4.30 to Equation 4.32 with $M = M_I$ results in

$$\frac{\Delta M_I}{M_I} \approx \frac{1 + F_{I_{\text{nom}}}}{A_{\text{nom}} F_{I_{\text{nom}}}} \frac{\Delta A}{A_{\text{nom}}} + \frac{\Delta R_F}{R_{F_{\text{nom}}}} - \frac{\Delta R_I}{R_{I_{\text{nom}}}}. \quad (4.35)$$

Again, it is of importance to find the worst case $\Delta M_I / M_I$, which is the maximum of its absolute value:

$$\text{Max} \left| \frac{\Delta M_I}{M_I} \right| \approx \left| \frac{1 + F_{I_{\text{nom}}}}{A_{\text{nom}} F_{I_{\text{nom}}}} \frac{\Delta A}{A_{\text{nom}}} \right| + \left| \frac{\Delta R_F}{R_{F_{\text{nom}}}} \right| + \left| \frac{\Delta R_I}{R_{I_{\text{nom}}}} \right|. \quad (4.36)$$

PROBLEMS

1. Retain the second-order term in the expansion of Equation 4.6 and show that the fractional error of Equation 4.9 is less than $|\Delta A/A_{\text{nom}}|$.

2. A noninverting amplifier circuit with negative feedback uses an operational amplifier with an amplification of $A = 20,000 \pm 10\%$. Resistor values are $R_I = 200\,\Omega$ and $R_F = 1800\,\Omega$. Find the nominal value of the resulting amplification M_N, and its error resulting from the 10% change in A.

3. Derive Equations 4.9, 4.11, 4.20, 4.23, 4.26, and 4.29.

4. Apply Equation 4.10 to the case of the voltage follower circuit. What is the fractional error in the resulting amplification M_N, if the operational amplifier used in the circuit has an amplification of $A = 10,000 \pm 10\%$?

5. In the noninverting amplifier circuit of Fig. 3.1, $A = 20,000 \pm 20\%$, $R_I = 100\,\Omega \pm 1\%$, and $R_F = 10,000\,\Omega \pm 1\%$. Find the nominal value of the resulting amplification M_N, and its worst case fractional error in percents.

6. In the noninverting amplifier circuit of Fig. 3.1, $A = 20,000 \pm 20\%$, $R_I = 100\,\Omega \pm 0.1\%$, and $R_F = 10,000\,\Omega \pm 0.1\%$. Find the nominal value of the resulting amplification M_N, and its worst case fractional error in percents. Compare the result to that of the preceding problem.

7. In the noninverting amplifier circuit of Fig. 3.1, $A = 10,000 \pm 20\%$, $R_F = 10,000\,\Omega \pm 1\%$, and input resistor R_I is adjustable between its minimum of $R_{I_{\min}}$ and its maximum of $R_{I_{\max}}$. Find the maximum value of $R_{I_{\min}}$ and the minimum value of $R_{I_{\max}}$ thus that M_N can be always adjusted to be equal to 100.

8. In the inverting amplifier circuit of Fig. 3.2, $A = 100 \pm 10\%$, $R_I = 100\,\Omega \pm 1\%$, and $R_F = 10,000\,\Omega \pm 1\%$. Find the nominal value of the resulting amplification M_I, and its worst case fractional error in percents.

9. In the inverting amplifier circuit of Fig. 3.2, $A = 10,000$ and $R_I = 101\,\Omega$. Find the value of the feedback resistor R_F such that the resulting amplification $M_I = 100$.

10. In the noninverting amplifier circuit of Fig. 3.1, $A = 10,000$, $R_I = 100 \, \Omega$, and $R_F = 10,000 \, \Omega$. Find the value of $\Delta M_N / M_N$ if the values of both R_I and R_F increase by 10%.

11. Consider the compound differential amplifier circuit of Fig. 3.8 with components and input voltages as given in Problem 15 of Chapter 3. Find the worst case fractional change in the output voltage V_{out}, if all resistors can vary by as much as $\pm 1\%$ each.

12. In the amplifier circuit with positive feedback shown in Fig. 3.9, $A = 100 \pm 0.1\%$, $R_I = 100 \, \Omega$, and $R_F = 10,000 \, \Omega$. Estimate the value of $|\Delta M/M|$ resulting from the 0.1% change in A.

13. Utilize Equations 3.6, 4.1, 4.2, 4.3, and 4.4, with $M = M_N$, and derive Equation 4.9 without using Taylor series expansion.

14. Utilize Equations 3.2, 3.6, 4.2, 4.14, 4.15, and 4.17, with $M = M_N$, and derive Equation 4.20 without using Taylor series expansion.

5

oooooooooooooooooooooooooooooooo

Transient Response and Frequency Response of Operational Amplifiers

Thus far it has been assumed that the amplification (differential voltage amplification) of the operational amplifier was a real positive number. This assumption, while it has led to significant results, becomes untenable when transient response and frequency response have to be determined, since a real amplification A would entail infinite bandwidths, zero rise times, and zero delay times— all physically impossible. In this chapter, representations of A suitable for use in the determination of transient response and of bandwidth will be examined.

LAG NETWORKS

The *lag network* of Fig. 5.1 often provides a reasonable approximation to an amplifier stage in an operational amplifier.* Variables I_{in} and V_{out} are, respectively, a current signal such as the collector

---

* An alternate configuration of the lag network is given in Figure 5.11, page 64.

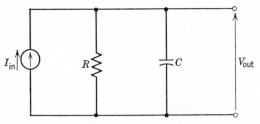

Figure 5.1. A lag network.

current output of a transistor stage, and a voltage signal such as the voltage input to the next stage.

If V_{out} and I_{in} are represented in the time domain, that is, $V_{out} = V_{out}(t)$ and $I_{in} = I_{in}(t)$, then it can be shown that the ratio of their Laplace transforms

$$\frac{\mathscr{L}\{V_{out}(t)\}}{\mathscr{L}\{I_{in}(t)\}} = \frac{R}{1 + RCs}, \tag{5.1}$$

where s is the Laplace transform variable in the transform domain.*

Example 5.1. In the circuit of Fig. 5.1, $R = 1000\ \Omega$ and current I_{in} is a function of time as $I_{in}(t) = 1\ \text{mA} \times u(t)$, i.e., it is a step-function with a magnitude of 1 mA. The Laplace transform of $I_{in}(t)$ is

$$\mathscr{L}\{I_{in}(t)\} = \mathscr{L}\{1\ \text{mA} \times u(t)\} = 1\ \text{mA} \times \mathscr{L}\{u(t)\} = \frac{1\ \text{mA}}{s}.$$

Thus, the Laplace transform of voltage $V_{out}(t)$ is given by

$$\mathscr{L}\{V_{out}(t)\} = \frac{R}{1 + RCs}\ \mathscr{L}\{I_{in}(t)\}$$

$$= \frac{1\ \text{mA}}{s}\ \frac{1000\ \Omega}{1 + RCs} = \frac{1\ \text{V}}{s(1 + RCs)}.$$

The inverse Laplace transform of this is

$$V_{out}(t) = \mathscr{L}^{-1}\left\{\frac{1\ \text{V}}{s(1 + RCs)}\right\} = 1\ \text{V} \times (1 - e^{-t/RC}).$$

* For a summary of the use of Laplace transforms in electronic circuits, see, e.g., A. Barna, *High-Speed Pulse Circuits*, Wiley-Interscience, New York, 1970.

If, alternatively, I_{in} and V_{out} are decomposed into sinusoidal waveforms in the frequency domain, then for each component at angular frequency ω the transfer function can be written as

$$\frac{V_{out}(\omega)}{I_{in}(\omega)} = \frac{R}{1 + j\omega RC} , \qquad (5.2)$$

where angular frequency ω is related to (cyclic) frequency f as

$$\omega = 2\pi f. \qquad (5.3)$$

By defining a *corner frequency* f_0 as

$$f_0 \equiv \frac{1}{2\pi RC} , \qquad (5.4)$$

the transfer function of Equation 5.2 can be written as

$$\frac{V_{out}(f)}{I_{in}(f)} = \frac{R}{1 + jf/f_0} . \qquad (5.5)$$

Example 5.2. In the circuit of Fig. 5.1, I_{in} is a sinusoidal waveform with a frequency of $f = 10$ MHz; $R = 10,000\ \Omega$, and $C = 10$ pF. Thus, from Equation 5.4,

$$f_0 = \frac{1}{2\pi RC} = \frac{1}{2\pi 10,000 \times 10^{-11}} = 1.59 \text{ MHz} ,$$

and the transfer function of Equation 5.5 becomes

$$\frac{V_{out}(f)}{I_{in}(f)} = \frac{10,000\ \Omega}{1 + j\,10 \text{ MHz}/1.59 \text{ MHz}} = \frac{10,000\ \Omega}{1 + 6.28\,j} .$$

It is of interest to determine the magnitude and the phase of the transfer function. From Equation 5.5, the magnitude is

$$\left| \frac{V_{out}(f)}{I_{in}(f)} \right| = \left| \frac{R}{1 + jf/f_0} \right| = \frac{R}{\sqrt{1 + (f/f_0)^2}}, \qquad (5.6)$$

and the phase φ is

$$\varphi \equiv \left/ \frac{V_{out}(f)}{I_{in}(f)} \right. = \left/ \frac{R}{1 + jf/f_0} \right.$$

$$= -\arctan(f/f_0) = -\frac{\pi}{2} + \arctan(f_0/f). \qquad (5.7)$$

Example 5.3. In the circuit of Fig. 5.1, $R = 10,000\ \Omega$, $f = 10$ MHz, and $f_0 = 1.59$ MHz. By utilizing Equation 5.6, the magnitude of the transfer function is

$$\left| \frac{V_{\text{out}}(f)}{I_{\text{in}}(f)} \right| = \frac{R}{\sqrt{1 + (f/f_0)^2}}$$

$$= \frac{10,000\ \Omega}{\sqrt{1 + (10\ \text{MHz}/1.59\ \text{MHz})^2}} = 1570\ \Omega.$$

Thus, if $I_{\text{in}}(f)$ has a magnitude of 1 mA, then the magnitude of $V_{\text{out}}(f)$ will be $1\ \text{mA} \times 1570\ \Omega = 1.57$ V. The phase φ, from Equation 5.7, is

$$\varphi = -\arctan (f/f_0) = -\arctan (10\ \text{MHz}/1.59\ \text{MHz})$$
$$= -81°.$$

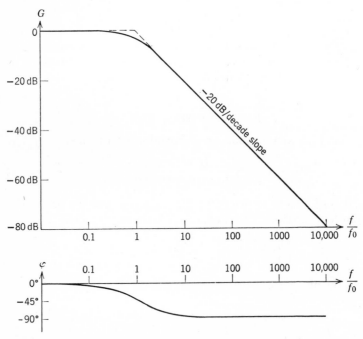

Figure 5.2. Bode plots $G \equiv 20\ \text{dB}\ \log_{10} [1 + (f/f_0)^2]^{-\frac{1}{2}}$ and $\varphi \equiv \underline{/(1 + jf/f_0)^{-1}}$ of a lag network.

It is customary to plot the *gain G of the lag network*, that can be defined as

$$G \equiv 20\,\mathrm{dB} \times \log_{10} \frac{1}{\sqrt{1 + (f/f_0)^2}},$$

where G is in *decibels* (dB) as shown in the upper part of Fig. 5.2. This representation, known as the *Bode plot of the gain*, shows that at $f/f_0 = 1$ the gain is down by 3 decibels from its value at $f = 0$.* Also, for $f \gg f_0$, the gain drops at a rate of -20 dB/decade of frequency, or approximately -6 dB/octave. In many cases the gain can be approximated by the two straight lines of $G = G(f = 0)$ for $f \ll f_0$ and $G = G(f = 0) - 20\,\mathrm{dB}\,\log_{10}(f/f_0)$ for $f \gg f_0$; these two straight lines intersect at the frequency of $f = f_0$; hence, the name corner frequency. This *piecewise linear approximation* of the gain is shown in Fig. 5.2 by broken lines. The *Bode plot of the phase* φ as function of frequency is also given in the figure.

For certain ranges of f/f_0, Equations 5.6 and 5.7 can be approximated by simpler expressions. For small values of f/f_0, several approximations of the magnitude given by Equation 5.6 are possible. One of these is a binomial expansion:

$$\left| \frac{V_{\mathrm{out}}(f)}{I_{\mathrm{in}}(f)} \right| \approx R\left[1 - \frac{1}{2}\left(\frac{f}{f_0}\right)^2 \right], \qquad (f/f_0)^2 \ll 1. \qquad (5.8)$$

Another approximation of Equation 5.6 for small values of f/f_0 results from a logarithmic expansion:

$$\left| \frac{V_{\mathrm{out}}(f)}{I_{\mathrm{in}}(f)} \right| \approx R\,\exp\left[-\frac{1}{2}\left(\frac{f}{f_0}\right)^2 \right], \qquad (f/f_0)^2 \ll 1. \qquad (5.9)$$

For large values of f/f_0, Equation 5.6 can be approximated as

$$\left| \frac{V_{\mathrm{out}}(f)}{I_{\mathrm{in}}(f)} \right| \approx R\frac{f_0}{f}, \qquad (f/f_0)^2 \gg 1. \qquad (5.10)$$

The phase φ of Equation 5.7 can be approximated for small values of f/f_0 as

$$\varphi \approx -\frac{f}{f_0}, \qquad (f/f_0)^2 \ll 1, \qquad (5.11)$$

* H. W. Bode, *Network Analysis and Feedback Amplifier Design*, D. Van Nostrand Company, Princeton, 1945.

and for large values of f/f_0 as

$$\varphi \approx -\frac{\pi}{2} + \frac{f_0}{f}, \qquad (f/f_0)^2 \gg 1. \qquad (5.12)$$

Example 5.4. In the circuit of Fig. 5.1, $f_0 = 1/2\pi RC = 1$ MHz. At a frequency of $f = 0.1$ MHz, $f/f_0 = 0.1$. A first approximation of the magnitude of the transfer function, from Equation 5.8, is

$$\left| \frac{V_{\text{out}}(f)}{I_{\text{in}}(f)} \right| \approx R.$$

A better approximation from Equation 5.8 would be

$$\left| \frac{V_{\text{out}}(f)}{I_{\text{in}}(f)} \right| \approx R\left[1 - \frac{1}{2}\left(\frac{f}{f_0}\right)^2 \right] = R[1 - \tfrac{1}{2}0.1^2] = 0.995R,$$

or from Equation 5.9

$$\left| \frac{V_{\text{out}}(f)}{I_{\text{in}}(f)} \right| \approx R \exp\left[-\frac{1}{2}\left(\frac{f}{f_0}\right)^2 \right]$$

$$= R \exp\left[-\tfrac{1}{2}0.1^2\right] = 0.99501R.$$

The exact magnitude is given by Equation 5.6 as

$$\left| \frac{V_{\text{out}}(f)}{I_{\text{in}}(f)} \right| = \frac{R}{\sqrt{1 + (f/f_0)^2}} = \frac{R}{\sqrt{1 + 0.1^2}} = 0.995037R.$$

The value of the phase φ can be approximated from Equation 5.11 as

$$\varphi \approx -\frac{f}{f_0} = -0.1 = -5.73°.$$

The exact value of the phase φ, from Equation 5.7, is

$$\varphi = -\arctan(f/f_0) = -\arctan(0.1) = -5.71°.$$

CASCADED LAG NETWORKS

The lag network of Fig. 5.1 in many cases provides a reasonable approximation of a stage in an operational amplifier. Usually, however, there are several stages that have to be represented by several lag networks. Fortunately, in most cases these are separated and the resulting transfer function can be written as a product of transfer functions. Thus, in the frequency domain, the transfer function of a two-stage amplifier can be written in the form

$$\text{constant} \times \frac{1}{1 + jf/f_1} \, \frac{1}{1 + jf/f_2}.$$

In Fig. 5.1, a current input and a voltage output were assumed. In the case when an entire operational amplifier is to be represented, both the input and the output signals are voltages. Hence, the amplification A of an operational amplifier consisting of two stages can be written as

$$A = \frac{A_{DC}}{(1 + jf/f_1)(1 + jf/f_2)}, \tag{5.13}$$

where $A_{DC} \equiv A(f = 0)$ is the amplification of the operational amplifier at zero frequency: a real positive dimensionless number.

Example 5.5. A two-stage operational amplifier has an amplification at zero frequency of $A_{DC} = 1000$ and it can be represented by two separated lag networks. One of these consists of a 100,000-Ω resistance and a 5-pF capacitance, the other one of a 1000-Ω resistance and a 5-pF capacitance. Since the designations of f_1 and f_2 are interchangeable, one can write

$$f_1 = \frac{1}{2\pi \, 100,000 \times 5 \times 10^{-12}} = 318 \text{ kHz}$$

and

$$f_2 = \frac{1}{2\pi \, 1000 \times 5 \times 10^{-12}} = 31.8 \text{ MHz}.$$

Thus, amplification A becomes

$$A = \frac{1000}{(1 + jf/318 \text{ kHz})(1 + jf/31.8 \text{ MHz})}.$$

In general, the *gain* is defined as $G \equiv 20 \text{ dB} \times \log_{10} |A|$, and the phase is $\varphi = \underline{/A}$. The gain and the phase plots of a two-stage amplifier described by Equation 5.13 with $f_2/f_1 = 100$ and $A_{DC} = 1000$ are shown in Fig. 5.3. Since the logarithm of a product is the sum of the logarithms, the gain is obtained as the sum of the individual gains of the two stages; the individual phases of the two stages are added linearly to obtain the resulting phase. It can be seen that for $f \gg f_2 \geq f_1$ the gain falls off at a rate of -40 dB/decade of frequency.

Figure 5.4 shows the special case of two equal corner frequencies $f_1 = f_2$, that is, the case of

$$A = \frac{A_{DC}}{(1 + jf/f_1)^2}.$$ (5.14)

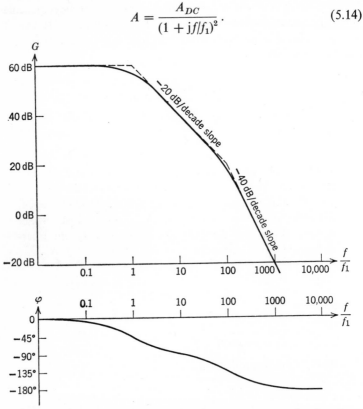

Figure 5.3. Bode plots $G \equiv 20$ dB $\log_{10} |A|$ and $\varphi \equiv \underline{/A}$ for the two-stage operational amplifier of Equation 5.13 with $A_{DC} = 1000$ and $f_2 = 100 f_1$.

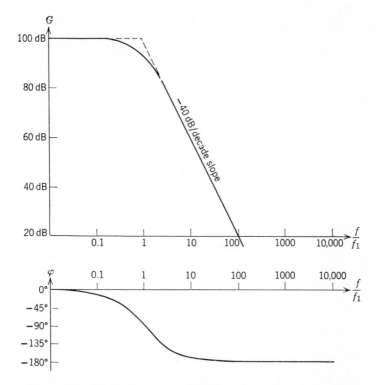

Figure 5.4. Bode plots $G \equiv 20 \text{ dB} \log_{10} |A|$ and $\varphi = \underline{/A}$ for an operational amplifier with $A_{DC} = 100,000$ consisting of two identical stages.

For a three-stage amplifier, Equation 5.13 can be extended to

$$A = \frac{A_{DC}}{(1 + jf/f_1)(1 + jf/f_2)(1 + jf/f_3)}, \qquad (5.15)$$

and to similar expressions for larger number of stages.

Example 5.6. In the expression of Equation 5.15, $f_1 = 1$ MHz, $f_2 = 4$ MHz, $f_3 = 40$ MHz, and $A_{DC} = 10,000$. What is the magnitude $|A|$ and the phase $\underline{/A}$ at a frequency of $f = 0.1$ MHz?

The magnitude of A,

$$|A| = \frac{|A_{DC}|}{|1 + jf/f_1| \, |1 + jf/f_2| \, |1 + jf/f_3|} .$$

By utilizing Equations 5.6 and 5.8,

$$\frac{1}{|1 + jf/f_1|} \approx 1 - \frac{1}{2}\left(\frac{f}{f_1}\right)^2 = 1 - \tfrac{1}{2}0.1^2 = 0.995,$$

$$\frac{1}{|1 + jf/f_2|} \approx 1 - \frac{1}{2}\left(\frac{f}{f_2}\right)^2 = 1 - \frac{1}{2}\left(\frac{0.1}{4}\right)^2 = 0.9997,$$

and

$$\frac{1}{|1 + jf/f_3|} \approx 1 - \frac{1}{2}\left(\frac{f}{f_3}\right)^2 = 1 - \frac{1}{2}\left(\frac{0.1}{40}\right)^2 = 0.999997;$$

thus, $|A| \approx 10,000 \times 0.995 \times 0.9997 \times 0.999997 = 9947$.

The phase can be obtained by adding the individual phases:

$$\underline{/A} = \underline{/A_{DC}} + \left/\frac{1}{1 + jf/f_1}\right. + \left/\frac{1}{1 + jf/f_2}\right. + \left/\frac{1}{1 + jf/f_3}\right. .$$

By applying Equation 5.7, this can be written as

$$\underline{/A} = -\arctan(f/f_1) - \arctan(f/f_2) - \arctan(f/f_3),$$

which, by utilizing the approximation of Equation 5.11, becomes

$$\underline{/A} \approx -\frac{f}{f_1} - \frac{f}{f_2} - \frac{f}{f_3} = -\frac{0.1}{1} - \frac{0.1}{4} - \frac{0.1}{40}$$
$$= -0.1275 = -7.3°.$$

MODIFIED LAG NETWORKS

Two equivalent forms of a *modified lag network* are shown in Fig. 5.5. Such networks are of use in modifying the properties of operational amplifiers; usually they are not part of the amplifier, but are added externally. If V_{in} and V_{out} are represented in the time domain, then it can be shown that for the circuit of Fig. 5.5b,

$$\frac{\mathscr{L}\{V_{out}(t)\}}{\mathscr{L}\{V_{in}(t)\}} = \frac{1 + sR_2C}{1 + s(R_1 + R_2)C} . \qquad (5.16)$$

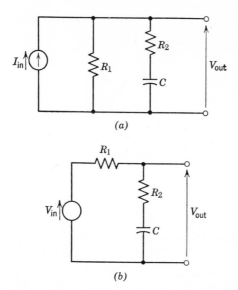

Figure 5.5. Two equivalent forms of a modified lag network.

If V_{in} and V_{out} are represented in the frequency domain, then the transfer function is

$$\frac{V_{out}(f)}{V_{in}(f)} = \frac{1 + jf/f_2}{1 + jf/f_1}, \tag{5.17a}$$

where corner frequencies f_1 and f_2 are defined by

$$f_1 \equiv \frac{1}{2\pi(R_1 + R_2)C} \tag{5.17b}$$

and

$$f_2 \equiv \frac{1}{2\pi R_2 C}. \tag{5.17c}$$

Bode plots of a modified lag network with $f_2 = 100f_1$ are shown in Fig. 5.6.

It can be seen from Equations 5.17b and 5.17c that $f_2 \geq f_1$; thus, the magnitude of Equation 5.17a,

$$\left| \frac{V_{out}(f)}{V_{in}(f)} \right| = \frac{|1 + jf/f_2|}{|1 + jf/f_1|} \tag{5.18}$$

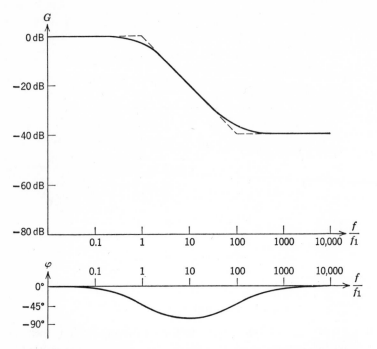

Figure 5.6.   Bode plots of a modified lag network with $f_2 = 100f_1$.

is monotonically decreasing as a function of frequency f. The phase can be written as

$$\underline{\bigg/ \frac{V_{out}(f)}{V_{in}(f)}} = \underline{\bigg/ \frac{1 + jf/f_2}{1 + jf/f_1}} = \underline{\big| 1 + jf/f_2} - \underline{\big| 1 + jf/f_1}$$

$$= \arctan\left(\frac{f}{f_2}\right) - \arctan\left(\frac{f}{f_1}\right). \qquad (5.19)$$

The ratio of f_2 and f_1 can be expressed from Equation 5.17 as

$$\frac{f_2}{f_1} = \frac{R_1 + R_2}{R_2} = 1 + \frac{R_1}{R_2}. \qquad (5.20)$$

Thus, if f_2/f_1 is given, it determines R_1/R_2, but C and one of R_1 and R_2 can be still chosen.

Example 5.7. In the modified lag network of Fig. 5.5, $(R_1 + R_2)/R_2 = 10$; thus $f_2/f_1 = 10$. The magnitude of the transfer function, from Equation 5.18, is

$$\left| \frac{V_{\text{out}}(f)}{V_{\text{in}}(f)} \right| = \frac{|1 + jf/f_2|}{|1 + jf/f_1|} = \frac{|1 + jf/(10f_1)|}{|1 + jf/f_1|}$$

and the phase φ of the transfer function becomes

$$\varphi = \left/ \frac{1 + jf/f_2}{1 + jf/f_1} \right. = \arctan\left(\frac{f}{f_2}\right) - \arctan\left(\frac{f}{f_1}\right)$$

$$= \arctan\left(\frac{f}{10f_1}\right) - \arctan\left(\frac{f}{f_1}\right).$$

If, alternatively, f_1 and f_2 are given, then R_1 and R_2 can be computed from

$$R_1 = \frac{1}{2\pi C}\left(\frac{1}{f_1} - \frac{1}{f_2}\right) \tag{5.21a}$$

and

$$R_2 = \frac{1}{2\pi C}\frac{1}{f_2}. \tag{5.21b}$$

LEAD NETWORKS

A *lead network* is shown in Fig. 5.7. As is the case for the modified lag network, the lead network has its utility in modifying the response of operational amplifiers and is usually located external to

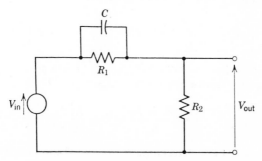

Figure 5.7. A lead network.

the amplifier. When V_{in} and V_{out} are represented in the time domain,

$$\frac{\mathscr{L}\{V_{out}(t)\}}{\mathscr{L}\{V_{in}(t)\}} = \frac{R_2}{R_1 + R_2} \frac{1 + sCR_1}{1 + sC\dfrac{R_1R_2}{R_1 + R_2}}. \tag{5.22}$$

In the frequency domain representation, the transfer function can be written as

$$\frac{V_{out}(f)}{V_{in}(f)} = \frac{f_1}{f_2} \frac{1 + jf/f_1}{1 + jf/f_2}, \tag{5.23a}$$

where corner frequencies f_1 and f_2 are defined by

$$f_1 \equiv \frac{1}{2\pi R_1 C} \tag{5.23b}$$

and

$$f_2 \equiv \frac{1}{2\pi C \dfrac{R_1R_2}{R_1 + R_2}}. \tag{5.23c}$$

It is seen from Equations 5.23b and 5.23c that $f_2 \geq f_1$; thus, the magnitude of Equation 5.23a,

$$\left| \frac{V_{out}(f)}{V_{in}(f)} \right| = \frac{f_1}{f_2} \frac{|1 + jf/f_1|}{|1 + jf/f_2|} \tag{5.24}$$

is monotonically increasing as function of frequency f. The phase φ of Equation 5.23a can be written as

$$\varphi = \left/ \frac{1 + jf/f_1}{1 + jf/f_2} \right. = \arctan\left(\frac{f}{f_1}\right) - \arctan\left(\frac{f}{f_2}\right). \tag{5.25}$$

Bode plots of a lead network with $f_2 = 100f_1$ are shown in Fig. 5.8. It can be seen from Equation 5.23 that

$$\frac{f_2}{f_1} = \frac{R_1 + R_2}{R_2} = 1 + \frac{R_1}{R_2}. \tag{5.26}$$

Thus if f_2/f_1 is given, it determines R_1/R_2, but C and one of R_1 and R_2 can be still chosen. If f_1 and f_2 are given, then R_1 and R_2 can be

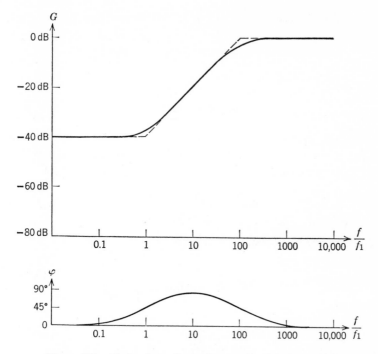

Figure 5.8. Bode plots of a lead network with $f_2 = 100f_1$.

computed from

$$R_1 = \frac{1}{2\pi C} \frac{1}{f_1} \qquad (5.27a)$$

and

$$R_2 = \frac{1}{2\pi C} \frac{1}{f_2 - f_1}. \qquad (5.27b)$$

Example 5.8. In the lead network of Fig. 5.7, $(R_1 + R_2)/R_2 = 10$; thus $f_2/f_1 = 10$. The magnitude of the transfer function, from Equation 5.24, is

$$\left| \frac{V_{\text{out}}(f)}{V_{\text{in}}(f)} \right| = \frac{f_1}{f_2} \frac{|1 + jf/f_1|}{|1 + jf/f_2|} = \frac{1}{10} \frac{|1 + jf/f_1|}{|1 + jf/10f_1|}$$

and the phase φ of the transfer function is

$$\varphi = \left/\frac{1 + jf/f_1}{1 + jf/f_2}\right. = \arctan\left(\frac{f}{f_1}\right) - \arctan\left(\frac{f}{10f_1}\right).$$

VOLTAGE DIVIDER NETWORKS

A *voltage divider network* consisting of two resistances and two capacitances is shown in Fig. 5.9. When V_{in} and V_{out} are represented in the frequency domain, it can be shown that the transfer function is

$$\frac{V_{out}(f)}{V_{in}(f)} = \frac{R_P}{R_S + R_P}\frac{1 + jf/f_1}{1 + jf/f_2}, \tag{5.28a}$$

where corner frequencies f_1 and f_2 are defined as

$$f_1 \equiv \frac{1}{2\pi R_S C_S} \tag{5.28b}$$

and

$$f_2 = \frac{1}{2\pi \dfrac{R_S R_P}{R_S + R_P}(C_S + C_P)}. \tag{5.28c}$$

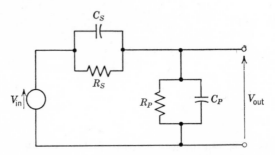

Figure 5.9. A voltage divider network.

Example 5.9. In the voltage divider network of Fig. 5.9, $R_S = 10,000\,\Omega$, $R_P = 100\,\Omega$, $C_S = 0.5\,\text{pF}$, and $C_P = 5\,\text{pF}$.

Thus,

$$f_1 \equiv \frac{1}{2\pi R_S C_S} = \frac{1}{2\pi\, 10{,}000 \times 0.5 \times 10^{-12}} = 31.8 \text{ MHz},$$

$$f_2 \equiv \frac{1}{2\pi\, \dfrac{R_S R_P}{R_S + R_P}\,(C_S + C_P)} = 289 \text{ MHz},$$

and the transfer function becomes

$$\frac{V_{\text{out}}(f)}{V_{\text{in}}(f)} = \frac{R_P}{R_S + R_P}\, \frac{1 + jf/f_1}{1 + jf/f_2} = 0.0099\, \frac{1 + jf/31.8 \text{ MHz}}{1 + jf/289 \text{ MHz}}.$$

Limiting cases of the voltage divider network are shown in Fig. 5.10. The first one (Fig. 5.10a) is a resistive divider with a frequency-independent $V_{\text{out}}/V_{\text{in}} = R_P/(R_P + R_S)$. The second one (Fig. 5.10b) has a transfer function of

$$\frac{V_{\text{out}}(f)}{V_{\text{in}}(f)} = \frac{R_P}{R_P + R_S}\, \frac{1}{1 + jf/f_0}, \qquad (5.29a)$$

where corner frequency f_0 is

$$f_0 \equiv \frac{R_P + R_S}{2\pi C R_P R_S}, \qquad (5.29b)$$

that is, this circuit has the frequency characteristics of a lag network. The circuit of Fig. 5.10c is a lead network (see Equation 5.23).

The last circuit (Fig. 5.10d) is a *compensated voltage divider network* where

$$R_S C_S = R_P C_P. \qquad (5.30)$$

It can be shown that when Equation 5.30 is satisfied, the transfer function of Equation 5.28a reduces to

$$\frac{V_{\text{out}}(f)}{V_{\text{in}}(f)} = \frac{R_P}{R_S + R_P}, \qquad (5.31)$$

independent of frequency and identical to the transfer function of the resistive divider of Fig. 5.10a. Thus, if the frequency-independent transfer function of the divider of Fig. 5.10a is desired but unavoidable

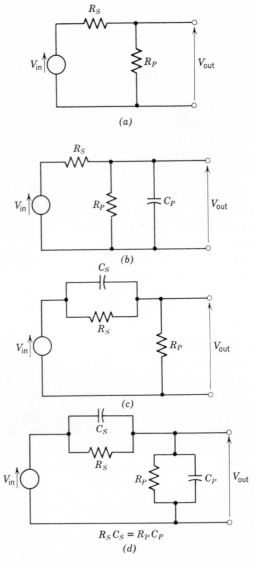

Figure 5.10. Limiting cases of the voltage divider network. (*a*) Resistive divider. (*b*) Lag network. (*c*) Lead network. (*d*) Compensated voltage divider network.

capacitances are present, one of the capacitances should be increased such that Equation 5.30 is satisfied.

Example 5.10. In the preceding example, $R_S = 10,000\ \Omega$, $R_P = 100\ \Omega$, $C_S = 0.5\ \text{pF}$, and $C_P = 5\ \text{pF}$ were given for the component values in the circuit of Fig. 5.9 and a frequency-dependent transfer function $V_{\text{out}}(f)/V_{\text{in}}(f)$ resulted. If a frequency-independent transfer function is desired, then, from Equation 5.30,

$$\frac{C_P}{C_S} = \frac{R_S}{R_P} = \frac{10,000\ \Omega}{100\ \Omega} = 100$$

should hold. In the original circuit, however,

$$\frac{C_P}{C_S} = \frac{5\ \text{pF}}{0.5\ \text{pF}} = 10.$$

A $C_P/C_S = 100$ can be attained by connecting an additional 45-pF capacitance parallel with the original C_P, resulting in a

$$\frac{C_P}{C_S} = \frac{5\ \text{pF} + 45\ \text{pF}}{0.5\ \text{pF}} = 100$$

and in a frequency-independent transfer function.

PROBLEMS

1. For the circuit of Fig. 5.1, sketch output voltage V_{out} as function of time, if input current I_{in} is a delta-function $I_{\text{in}}(t) = 10^{-10}$ coulomb $\times\ \delta(t)$; $R = 1000\ \Omega$, and $C = 10\ \text{pF}$.

2. Derive Equations 5.2 and 5.5.

3. Evaluate Equations 5.6 and 5.7 if, in the circuit of Fig. 5.1, $R = 2000\ \Omega$, $C = 20\ \text{pF}$, and current I_{in} is a sinewave with a frequency of 20 MHz.

4. Show that the phase plot of Fig. 5.2 is antisymmetric (invariant under 180° rotation) around the $f/f_0 = 1$, $\varphi = -45°$ point.

5. Derive Equations 5.6 and 5.7.

6. An alternate configuration of the lag network is shown in Fig. 5.11. Derive the expressions $\mathscr{L}\{V_{out}(t)\}/\mathscr{L}\{V_{in}(t)\}$ and $V_{out}(f)/V_{in}(f)$.

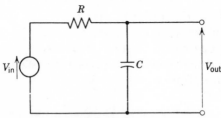

Figure 5.11.

7. Show that an approximation of the phase φ of Equation 5.7 for small values of f/f_0, which is more accurate than that of Equation 5.11, is given by

$$\varphi \approx -\frac{f}{f_0} + \frac{1}{3}\left(\frac{f}{f_0}\right)^3.$$

8. Show that an approximation of the phase φ of Equation 5.7 for large values of f/f_0, which is more accurate than that of Equation 5.12, is given by

$$\varphi \approx -\frac{\pi}{2} + \frac{f_0}{f} - \frac{1}{3}\left(\frac{f_0}{f}\right)^3.$$

Evaluate this expression, Equation 5.12, and Equation 5.7 for $f/f_0 = 10$. Present the results in degrees.

9. Show that the tangent drawn to the phase plot φ of Fig. 5.2 at $f/f_0 = 1$, $\varphi = -45°$ reaches $f/f_0 = 10$ at a phase of

$$\varphi = -45° - \frac{180° \ln 10}{2} \approx -45° - 66° = -111°.$$

10. Derive Equation 5.9 by expanding $\ln\left[1/\sqrt{1 + (f/f_0)^2}\right]$ into a Taylor series.

11. Show that the approximation of Equation 5.9 evaluated at $f/f_0 = 1$ would result in a gain that is down by $10\,dB\,\log_{10} e \approx 4.3\,dB$ from its value at $f = 0$.

12. A stage of an operational amplifier can be represented by the circuit of Fig. 5.1 with $R = 10,000\,\Omega$ and $C = 5\,pF$. At what frequency is the gain down by $3\,dB$ from its value at zero frequency?

13. A two-stage operational amplifier has an amplification at zero frequency of $A_{DC} = 10,000$. It consists of two identical stages, each of which can be represented by a lag network with $R = 10,000\ \Omega$ and $C = 5\ \text{pF}$. Sketch the Bode plots, using piecewise linear approximation for the gain.

14. Derive $\mathcal{L}\{V_{\text{out}}(t)\}/\mathcal{L}\{V_{\text{in}}(t)\}$ for the amplifier described in the preceding problem. Find the output voltage as function of time if the input voltage is a step-function $V_{\text{in}}(t) = 1\ \text{mV} \times u(t)$.

15. Use Equation 5.9 and show that for $f/f_1 \ll 1$ and $f/f_2 \ll 1$ the magnitude of Equation 5.13 can be approximated as

$$|A| \approx A_{DC} \exp\left\{ -\frac{1}{2}\left[\left(\frac{f}{f_1}\right)^2 + \left(\frac{f}{f_2}\right)^2 \right] \right\}.$$

16. A three-stage operational amplifier has an amplification in the form of Equation 5.15 with $A_{DC} = 1000$. The three stages are characterized by $f_1 = 1\ \text{MHz}$, $f_2 = 10\ \text{MHz}$, and $f_3 = 50\ \text{MHz}$. Plot the Bode plots of amplification A, using piecewise linear approximation for the gain.

17. Show that the two forms of the modified lag network shown in Fig. 5.5 are equivalent if $V_{\text{in}} = R_1 I_{\text{in}}$.

18. Derive Equations 5.16, 5.17, and 5.20.

19. A modified lag network consists of the circuit of Fig. 5.5b with $R_1 = 10,000\ \Omega$, $R_2 = 200\ \Omega$, and $C = 1000\ \text{pF}$. What are the values of f_1 and f_2 in Equation 5.17? Sketch the Bode plots, using piecewise linear approximation for the gain. Repeat for $R_1 = 5000\ \Omega$.

20. Derive Equations 5.22, 5.23, and 5.27.

21. A lead network consists of the circuit of Fig. 5.7 with $R_1 = 10,000\ \Omega$, $R_2 = 100\ \Omega$, and $C = 10\ \text{pF}$. What are the values of f_1 and f_2 in Equation 5.23? Sketch the Bode plots, using piecewise linear approximation for the gain. Repeat for $R_1 = 5000\ \Omega$.

22. In the voltage divider network of Fig. 5.9, $R_S = 10,000\ \Omega$, $R_P = 1000\ \Omega$, $C_S = 0.5\ \text{pF}$, and $C_P = 10\ \text{pF}$. What is $V_{\text{out}}(f)/V_{\text{in}}(f)$? Modify the circuit by the addition of a capacitance such that the transfer function $V_{\text{out}}(f)/V_{\text{in}}(f)$ is frequency independent. Give the location and the value of the capacitance required.

23. Derive $\mathscr{L}\{V_{\text{out}}(t)\}/\mathscr{L}\{V_{\text{in}}(t)\}$ in the voltage divider network of Fig. 5.9. Sketch output voltage $V_{\text{out}}(t)$ if the input voltage is a step-function $V_{\text{in}}(t) = 1 \text{ V} \times u(t)$; $R_S = 1000 \ \Omega$, $R_P = 2000 \ \Omega$, and $C_P = 1000 \text{ pF}$. Assume $C_S = 1000 \text{ pF}$, 2000 pF, and 5000 pF.

24. Sketch the Bode plots for Examples 5.7 and 5.8.

6

OOOOOOOOOOOOOOOOOOOOOOOOOOOOOOOOOOO

Frequency Response and Transient Response of Feedback Amplifier Circuits

Frequency response and transient response of operational amplifiers were discussed in the preceding chapter. The resulting frequency response and transient response of a circuit, however, can be altered by the application of feedback. In some cases, the application of feedback results in an unstable system, that is, in one that provides an output signal without the presence of an input signal. Stability conditions of feedback amplifier circuits will be discussed in the next chapter; here it will be assumed that the system is stable and attention will be focused on the discussion of the resulting frequency response and transient response.

FREQUENCY RESPONSE

In the case of a noninverting feedback amplifier circuit utilizing an operational amplifier that, in the frequency domain, has an amplification of

$$A = \frac{A_{DC}}{1 + \mathrm{j}f/f_0}, \qquad (6.1)$$

67

the resulting amplification, M_N, is

$$M_N = \frac{A}{1 + AF_N} , \qquad (6.2)$$

where feedback return F_N is a real positive dimensionless number. Substitution of Equation 6.1 into Equation 6.2 results in an expression for M_N that can be written as

$$M_N = \frac{M_{DC}}{1 + jf/[f_0(1 + A_{DC}F_N)]} , \qquad (6.3a)$$

or as

$$M_N = \frac{M_{DC}}{1 + jf/(f_0 A_{DC}/M_{DC})} , \qquad (6.3b)$$

where M_{DC} is the resulting amplification of the feedback amplifier circuit at zero frequency,

$$M_{DC} \equiv \frac{A_{DC}}{1 + A_{DC}F_N} . \qquad (6.3c)$$

Equation 6.3 is illustrated in Fig. 6.1 where the Bode plots are plotted for the case of $A_{DC} = 10,000$ with $M_{DC} = 10,000$ (i.e., $F_N = 0$), and with $M_{DC} = 100$ (i.e., $F_N = 0.0099$). It is seen that the corner

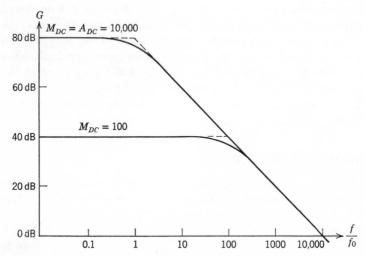

Figure 6.1. Bode plots of $|M_N|$ of Equation 6.3 for $A_{DC} = 10,000$ with $M_{DC} = 10,000$ and 100.

frequency B of $|M_N|$, $B = f_0 A_{DC}/M_{DC}$. For frequencies of $f \gg B$, $|M_N| \approx A_{DC} f_0/f$, independent of F_N and thus of M_{DC}. As a result, the gain plot of $|M_N|$ can be approximated by two straight lines: $|M_N| \approx M_{DC}$ for $f \ll B$ and $|M_N| \approx A_{DC} f_0/f$ for $f \gg B$. Here B equals the *3-dB bandwidth* (or simply *bandwidth*) of M_N: the frequency at which $|M_N|$ is down by 3 dB from its value at zero frequency.

Example 6.1. An operational amplifier is characterized by an amplification given by Equation 6.1 with $f_0 = 1$ MHz and $A_{DC} = 10,000$. It is used as a noninverting feedback amplifier with a resulting amplification at zero frequency of $M_{DC} = 200$. The resulting amplification M_N of the feedback amplifier circuit, from Equation 6.3*b*, is

$$M_N = \frac{M_{DC}}{1 + jf/(f_0 A_{DC}/M_{DC})} = \frac{200}{1 + jf/(1 \text{ MHz} \times 10,000/200)}$$
$$= \frac{200}{1 + jf/50 \text{ MHz}}.$$

Thus, the 3-dB bandwidth of the feedback amplifier circuit is $B = 50$ MHz.

In the case of an operational amplifier with an amplification of

$$A = \frac{A_{DC}}{(1 + jf/f_0)^2}, \qquad (6.4)$$

used in a noninverting feedback amplifier circuit, the resulting amplification M_N becomes

$$M_N = \frac{M_{DC}}{1 - \frac{M_{DC}}{A_{DC}} \left[\left(\frac{f}{f_0}\right)^2 - 2j\frac{f}{f_0} \right]}, \qquad (6.5a)$$

where M_{DC} is defined as

$$M_{DC} \equiv \frac{A_{DC}}{1 + A_{DC} F_N}, \qquad (6.5b)$$

and feedback return F_N is again a real positive dimensionless number.

Equation 6.5 is illustrated in Fig. 6.2 for $A_{DC} = 10,000$ with $M_{DC} = 10,000$ and 100. It can be shown that the corner frequency

B of $|M_N|$, $B = f_0\sqrt{A_{DC}/M_{DC}}$. For frequencies of $f \gg B$, $|M_N| \approx A_{DC}(f_0/f)^2$, independent of F_N and of M_{DC}. Figure 6.2 also shows that the shape of one of the gain plots is significantly different from the previous ones by having a resonance in the vicinity of $f \approx B$, returning to its zero frequency value in the vicinity of $f \approx B\sqrt{2}$. Except in the vicinity of such a resonance, the gain can be approximated by two straight lines determined by $|M_N| \approx M_{DC}$ for $f \ll B$ and $|M_N| \approx A_{DC}(f_0/f)^2$ for $f \gg B$. The 3-dB bandwidth of $|M_N|$ now depends on the details of the response, but it is still in the vicinity of B and it will be approximated by B in what follows.

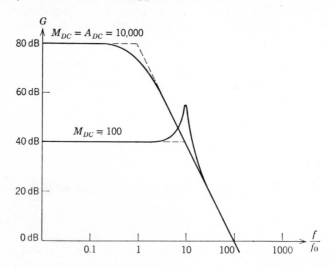

Figure 6.2. Bode plots of $|M_N|$ of Equation 6.5 for $A_{DC} = 10,000$ with $M_{DC} = 10,000$ and 100.

Example 6.2. An operational amplifier has an amplification A given by Equation 6.4 with $f_0 = 1$ MHz and $A_{DC} = 10,000$. It is used as a noninverting feedback amplifier with a resulting amplification at zero frequency of $M_{DC} = 200$. Thus, Equation 6.5a can be written as

$$M_N = \cfrac{200}{1 - \cfrac{200}{10,000}\left[\left(\cfrac{f}{1\text{ MHz}}\right)^2 - 2j\,\cfrac{f}{1\text{ MHz}}\right]}.$$

If

$$\left| 0.02\left[\left(\frac{f}{1 \text{ MHz}} \right)^2 - 2j\,\frac{f}{1 \text{ MHz}} \right] \right| \ll 1,$$

that is, if

$$\left(\frac{f}{1 \text{ MHz}} \right)^4 + 4\left(\frac{f}{1 \text{ MHz}} \right)^2 \ll 2500,$$

namely, $f \ll 7$ MHz, then $M_N \approx M_{DC} = 200$. If, alternatively, $(f/1 \text{ MHz})^4 + 4(f/1 \text{ MHz})^2 \gg 2500$, i.e., if $f \gg 7$ MHz, then it can be shown that

$$|M_N| \approx \frac{A_{DC}}{(f/1 \text{ MHz})^2} = \frac{10{,}000}{(f/1 \text{ MHz})^2} \approx \frac{200}{(f/7 \text{ MHz})^2}.$$

It can be also seen that the 3-dB bandwidth of the resulting feedback amplifier circuit is in the vicinity of $B = 7$ MHz.

TRANSIENT RESPONSE

When the input signal of an amplifier circuit is represented in the time domain, it is of interest to find the output signal as function of time. In the case when A has a single corner frequency f_0 as in Equation 6.1, the resulting amplification M_N has a single real corner frequency of $f_0 A_{DC}/M_{DC}$ (see Equation 6.3b). It can be shown that in this case the ratio of Laplace transforms,

$$\frac{\mathscr{L}\{V_{\text{out}}(t)\}}{\mathscr{L}\{V_{\text{in}}(t)\}} = \frac{M_{DC}}{1 + s\tau}, \qquad (6.6a)$$

where τ is defined as

$$\tau \equiv \frac{1}{2\pi B} \qquad (6.6b)$$

with B as the 3-dB bandwidth of the feedback amplifier circuit:

$$B \equiv f_0\,\frac{A_{DC}}{M_{DC}}. \qquad (6.6c)$$

Example 6.3. A noninverting feedback amplifier circuit utilizes a Type 741 operational amplifier.* Its amplification A can be approximated by Equation 6.1 with

* Properties of operational amplifier types used in the examples and problems are summarized in the Appendix.

$A_{DC} = 200,000$ and $f_0 = 10$ Hz; the feedback return is $F_N = 0.01$. Thus, the resulting amplification at zero frequency is

$$M_{DC} = \frac{A_{DC}}{1 + A_{DC}F_N} = \frac{200,000}{1 + 200,000 \times 0.01} \approx 100$$

and the resulting bandwidth is in the vicinity of

$$B = f_0 \frac{A_{DC}}{M_{DC}} = 10 \text{ Hz} \frac{200,000}{100} = 20 \text{ kHz};$$

hence

$$\tau = \frac{1}{2\pi B} = \frac{1}{2\pi \times 20 \text{ kHz}} \approx 8 \ \mu\text{s}.$$

If the input signal is a step-function with a magnitude of 1 mV, then

$$\mathscr{L}\{V_{\text{out}}(t)\} = \frac{1 \text{ mV}}{s} \frac{M_{DC}}{1 + s\tau} = \frac{1 \text{ mV}}{s} \frac{100}{1 + s \times 8 \ \mu\text{s}}.$$

Thus,

$$V_{\text{out}}(t) = 0.1 \text{ V} \times (1 - e^{-t/8 \ \mu\text{s}}).$$

In the case of Equation 6.4 where A has two identical corner frequencies f_0, it can be shown that the roots of the denominator of Equation 6.5a are at frequencies of $f = f_0(\text{j} \pm \sqrt{A_{DC}F_N})$, and that for $A_{DC}F_N \gg 1$ the output signal is composed of damped oscillatory terms of the form $e^{-\alpha t} \sin(\omega t + \varphi)$, where $\alpha \approx 2\pi f_0$. Furthermore, it can be shown that if the gain plot of $|M_N|$ vs frequency is approximated by two straight lines intersecting at a frequency $B = f_0\sqrt{A_{DC}/M_{DC}}$, then $\omega = 2\pi f_0\sqrt{A_{DC}F_N} \cong 2\pi B$.

Example 6.4. An operational amplifier has an amplification A given by Equation 6.4 with $f_0 = 1$ MHz and $A_{DC} = 10,000$. It is used as a noninverting feedback amplifier with a resulting amplification at zero frequency of $M_{DC} = 200$. It can be shown that with $s = \text{j}2\pi f$, the Laplace transform corresponding to Equation 6.5a can be written as

$$M_N(s) = \frac{4\pi^2 10^{16}}{(s + 2\pi 10^6 + \text{j}1.4\pi 10^7)(s + 2\pi 10^6 - \text{j}1.4\pi 10^7)}$$

From this, it can be shown by evaluating the inverse Laplace transform that the transient response is of the form $e^{-\alpha t} \sin(\omega t + \varphi)$ with $\alpha = 2\pi 10^6$ and $\omega = 1.4\pi 10^7$.

Similar approximations are applicable when there are two different corner frequencies in A. When A has more than two corner frequencies, such as in Equation 5.15, the resulting feedback amplifier circuit may not be stable, but it may oscillate. If, however, it is stable, then the output signal for a step-function input signal can be again composed of damped oscillatory terms of the form $e^{-\alpha t} \sin(\omega t + \varphi)$.

PROBLEMS

1. Derive Equation 6.3.

2. A Type 107 operational amplifier has an amplification A that can be approximated in the form of Equation 6.1 with $A_{DC} = 160,000$ and $f_0 = 5$ Hz. Sketch the Bode plots of the amplification A and of the resulting amplification M_N, if the amplifier is used as a non-inverting feedback amplifier with a resulting amplification at zero frequency of $M_{DC} = 10,000$. Compute the 3-dB bandwidth of $|M_N|$. Repeat for $M_{DC} = 100$ and for $M_{DC} = 10$.

3. A Type 9406 operational amplifier has an amplification A that can be approximated in the form of Equation 6.1 with $A_{DC} = 1000$ and $f_0 = 1.5$ MHz. Sketch the Bode plots of the amplification A and of the resulting amplification M_N, if the operational amplifier is used as a noninverting feedback amplifier with a resulting amplification at zero frequency of $M_{DC} = 10$. Calculate the 3-dB bandwidth of $|M_N|$. Repeat for $M_{DC} = 1$ (voltage follower).

4. Derive Equation 6.5.

5. Show that M_N of Equation 6.5a cannot become infinite.

6. Show that the magnitude of M_N calculated from Equation 6.5a has a maximum at a frequency of $f = f_0 \sqrt{A_{DC} F_N - 1}$ with a value of $0.5 \sqrt{A_{DC}/F_N}$.

7. A noninverting amplifier circuit with negative feedback utilizes a Type 9406 operational amplifier with an amplification A that can be approximated in the form of Equation 6.1 with $A_{DC} = 1000$ and $f_0 = 1.5$ MHz. Compute and sketch the output signal as function of time if the input signal is a step-function with a magnitude of 10 mV and if the resulting amplification at zero frequency is $M_{DC} = 100$.

8. A noninverting feedback amplifier circuit with negative feedback utilizes a Type 702A operational amplifier with an amplification A given by Equation 5.15 with $f_1 = 1$ MHz, $f_2 = 4$ MHz, $f_3 = 40$ MHz, and $A_{DC} = 4000$; feedback return $F_N = 0.01$. Assume that the resulting feedback amplifier circuit is stable (this will be proven in the next chapter). Sketch the Bode plots of $|A|$ and of $|M_N|$ using piecewise linear approximation. Estimate the 3-dB bandwidth of $|M_N|$ and the response time $1/\alpha$.

9. Derive the equivalent of Equation 6.3 for an inverting amplifier circuit with negative feedback. How does the result differ from Equation 6.3 if $F_I \ll 1$?

10. Derive the equivalent of Equation 6.5 for an inverting amplifier circuit with negative feedback. How does the result differ from Equation 6.5 if $F_I \ll 1$?

11. An operational amplifier has an amplification of $A = 400/(1 + jf/10 \text{ MHz})^3$ and it is used as a noninverting feedback amplifier with a feedback return of $F_N = 0.01$. Sketch the Bode plot of the gain for the resulting amplification M_N without using piecewise linear approximation. Repeat with $F_N = 0.02$.

12. Derive the results given in Example 6.4.

7

OOOOOOOOOOOOOOOOOOOOOOOOOOOOOOOO

Stability of Feedback Amplifier Circuits

Whether a noninverting feedback amplifier circuit characterized by a resulting amplification of

$$M_N = \frac{A}{1 + AF_N} \tag{7.1}$$

is stable for a given A and F_N is determined by the roots of $1 + AF_N$: if all roots have negative real parts, then the system is stable. It can be shown, although it will not be proven here, that an equivalent criterion of stability is the *Nyquist criterion* that can be expressed as follows: If in Equation 7.1 A describes a stable system, then the system described by M_N of Equation 7.1 is stable if, and only if, the line of AF_N plotted for $-\infty \leq f \leq +\infty$ in the complex plane, designated as *Nyquist diagram*, does not *encircle* the $-1 + j\,0$ point.* The case when A itself describes an unstable system will not be discussed here.

There can be several descriptions of the term *encircle*. For a physically realizable system $A(f = -\infty) = A(f = +\infty) = 0 + j\,0$, hence, the line of AF_N plotted from $f = -\infty$ to $f = +\infty$ is a closed curve. A topological interpretation pictures this closed curve as a

---

* N. Nyquist, "Regeneration Theory," *Bell System Tech. J.*, **11,** 126–147 (January 1932).

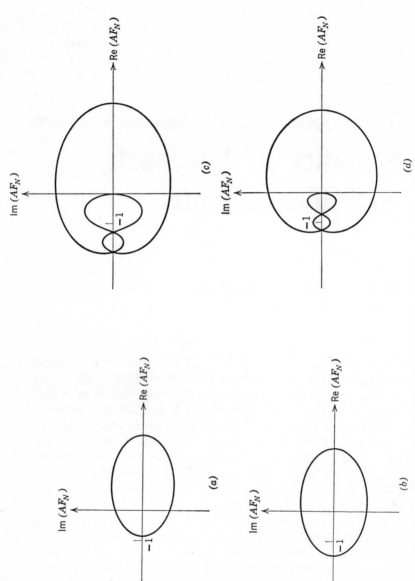

Figure 7.1 Nyquist diagrams of two stable systems, at (a) and (b), and of two unstable systems, at (c) and (d).

loop of string, with a stake driven into the complex plane at the $-1 + j\,0$ point. If the loop of string can be removed (without lifting it over the stake), then it does not encircle the stake and the system is stable. Thus, in Fig. 7.1a and in Fig. 7.1c the loop does not encircle the $-1 + j\,0$ point and the system is stable, while in Fig. 7.1b and in Fig. 7.1d the loop encircles the $-1 + j\,0$ point and the system is unstable. Alternatively, a vector can be drawn between the $-1 + j\,0$ point and a point on the line of AF_N in the complex plane. If the total angle traversed by this vector as it moves along the line of AF_N from $f = -\infty$ to $f = +\infty$ is zero, then the loop does not encircle the $-1 + j\,0$ point and the system is stable.

Example 7.1. An operational amplifier with an amplification of $A = A_{DC}/(1 + jf/f_0)$ is utilized as a noninverting feedback amplifier with a feedback return of F_N (a real positive number). At zero frequency, i.e., at $f = 0$, $AF_N = A_{DC}F_N$; at $f = \pm\infty$, $AF_N = 0$. The Nyquist diagram of AF_N is shown in Fig. 7.2. Since the phase of AF_N is

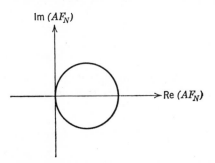

Figure 7.2. Nyquist diagram for Example 7.1.

always between $+90°$ and $-90°$, it never reaches $-180°$; thus, the $-1 + j\,0$ point cannot be encircled and the resulting feedback amplifier circuit is stable.

AMPLIFIER CIRCUITS CONSISTING OF LAG NETWORKS

Figure 7.3 illustrates the general characteristics of Nyquist diagrams corresponding to AF_N consisting of lag networks with transfer

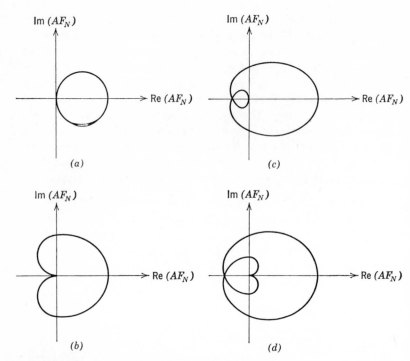

Figure 7.3. Nyquist diagrams of lag network responses. (*a*) One lag network. (*b*) Two lag networks. (*c*) Three lag networks. (*d*) Four lag networks.

functions of

$$AF_N = \frac{A_{DC}F_N}{1 + jf/f_1} \text{ (one lag network)}, \tag{7.2}$$

$$AF_N = \frac{A_{DC}F_N}{(1 + jf/f_1)(1 + jf/f_2)} \text{ (two lag networks)}, \tag{7.3}$$

$$AF_N = \frac{A_{DC}F_N}{(1 + jf/f_1)(1 + jf/f_2)(1 + jf/f_3)} \text{ (three lag networks)}, \tag{7.4}$$

$$AF_N = \frac{A_{DC}F_N}{(1 + jf/f_1)(1 + jf/f_2)(1 + jf/f_3)(1 + jf/f_4)}$$

$$\text{(four lag networks). } \tag{7.5}$$

It can be seen from Figure 7.3 that AF_N of Equations 7.2 and 7.3 always result in stable systems, while the stability of a system described by Equation 7.4 or 7.5 depends on the magnitude of $A_{DC}F_N$ and on the values of the corner frequencies.

The criterion of stability will be derived here for an amplifier circuit consisting of three lag networks, that is, for one characterized by Equation 7.4. Separating the real and imaginary parts in the denominator of Equation 7.4 results in

$$AF_N = \frac{A_{DC}F_N}{1 - \dfrac{f^2}{f_1 f_2} - \dfrac{f^2}{f_1 f_3} - \dfrac{f^2}{f_2 f_3} + j\left(\dfrac{f}{f_1} + \dfrac{f}{f_2} + \dfrac{f}{f_3} - \dfrac{f^3}{f_1 f_2 f_3}\right)}. \quad (7.6)$$

Since $A_{DC}F_N$ is real, Equation 7.6 is real if

$$\frac{f}{f_1} + \frac{f}{f_2} + \frac{f}{f_3} - \frac{f^3}{f_1 f_2 f_3} = 0. \quad (7.7)$$

One solution of Equation 7.7 is $f = 0$, corresponding to the intersection of the line of AF_N in Fig. 7.3c with the positive real axis. It is of interest here, however, to determine whether the intersection of the line with the negative real axis is to the right or to the left of the $-1 + j\,0$ point. If $f \neq 0$, then Equation 7.7 can be divided by f, resulting in

$$f = \pm\sqrt{f_1 f_2 + f_1 f_3 + f_2 f_3}. \quad (7.8)$$

These two frequencies, one positive and one negative, represent the point where the line of AF_N intersects the negative real axis in Fig. 7.3c. At these frequencies, AF_N of Equation 7.6 is real:

$$AF_N = \frac{A_{DC}F_N}{1 - \dfrac{f^2}{f_1 f_2} - \dfrac{f^2}{f_1 f_3} - \dfrac{f^2}{f_2 f_3}}. \quad (7.9)$$

Substituting Equation 7.8 into Equation 7.9 results in

$$AF_N = \frac{-A_{DC}F_N}{2 + \dfrac{f_2 + f_3}{f_1} + \dfrac{f_1 + f_3}{f_2} + \dfrac{f_1 + f_2}{f_3}}. \quad (7.10)$$

The system is stable if this is right of the $-1 + j\,0$ point, that is, if in Equation 7.10 $AF_N > -1$. This results in the criterion that the system is stable if

$$A_{DC}F_N < 2 + \frac{f_2 + f_3}{f_1} + \frac{f_1 + f_3}{f_2} + \frac{f_1 + f_2}{f_3}. \qquad (7.11)$$

Example 7.2.   A Type 702A operational amplifier is characterized by an amplification that can be described by three lag networks with $f_1 = 1$ MHz, $f_2 = 4$ MHz, and $f_3 = 40$ MHz. Thus, the criterion of stability becomes

$$A_{DC}F_N < 2 + \frac{4 \text{ MHz} + 40 \text{ MHz}}{1 \text{ MHz}} + \frac{1 \text{ MHz} + 40 \text{ MHz}}{4 \text{ MHz}}$$

$$+ \frac{1 \text{ MHz} + 4 \text{ MHz}}{40 \text{ MHz}} = 56.4.$$

The amplifier has a dc amplification of $A_{DC} = 4000$; hence, a stable system will result if feedback return

$$F_N < \frac{56.4}{A_{DC}} = \frac{56.4}{4000} = 0.014.$$

This criterion of stability can be also expressed by stating that in order to satisfy the inequality on feedback return F_N, it is required that M_{DC} be larger than

$$M_{DC} = \frac{A_{DC}}{1 + A_{DC}F_N} = \frac{4000}{1 + 56.4} = 70.$$

If M_{DC} is less than 70, then F_N is larger than 0.014 and the system is unstable. Therefore, it can be concluded that this configuration is not suitable for resulting amplifications of $M_{DC} < 70$, since the feedback amplifier circuit would oscillate at approximately the frequency given by Equation 7.8, i.e., at

$$f = \sqrt{f_1 f_2 + f_1 f_3 + f_2 f_3}$$
$$= \sqrt{1 \text{ MHz } 4 \text{ MHz} + 1 \text{ MHz } 40 \text{ MHz} + 4 \text{ MHz } 40 \text{ MHz}}$$
$$= 14.3 \text{ MHz}.$$

The exact frequency and amplitude of the oscillation would depend on the nonlinear properties of the operational amplifier and will not be analyzed here.

In the case of an inverting feedback amplifier circuit, by utilizing Equation 3.12, the resulting amplification

$$M_I = \frac{-A}{1 + (A + 1)F_I} = \frac{-1}{1 + F_I} \frac{A}{1 + \dfrac{AF_I}{1 + F_I}}, \qquad (7.12)$$

and the criterion of stability reduces to the case of the noninverting amplifier circuit if feedback return $F_I \ll 1$. If, however, F_I is not much less than 1, then it can be shown that the criterion becomes

$$\frac{A_{DC}F_I}{1 + F_I} < 2 + \frac{f_2 + f_3}{f_1} + \frac{f_1 + f_3}{f_2} + \frac{f_1 + f_2}{f_3}, \qquad (7.13)$$

that is, $A_{DC}F_I/(1 + F_I)$ has to be substituted in place of $A_{DC}F_N$.*

In the case when amplification A consists of four lag networks:

$$A = \frac{A_{DC}}{(1 + jf/f_1)(1 + jf/f_2)(1 + jf/f_3)(1 + jf/f_4)}, \qquad (7.14)$$

the criterion of stability will be derived for the limiting case when f_4 is small compared to $f_1, f_2,$ and f_3, that is, when

$$f_4 \ll f_1, \qquad (7.15a)$$

$$f_4 \ll f_2, \qquad (7.15b)$$

and

$$f_4 \ll f_3. \qquad (7.15c)$$

It can be shown that as a result of these approximations, in the vicinity where $\underline{/A} \approx -180°$ Equation 7.14 can be approximated as

$$A \approx \frac{A_{DC}}{(1 + jf/f_1)(1 + jf/f_2)(1 + jf/f_3)jf/f_4}. \qquad (7.16)$$

In order to determine if the feedback amplifier circuit is stable when negative feedback with feedback return F_N is applied, the value

---

* In fact, with this substitution, all stability criteria derived in this chapter for noninverting amplifiers can be applied to inverting amplifiers (see also Table 3).

of AF_N has to be found for $\underline{/AF_N} = -180°$. Utilizing Equation 7.16, AF_N can be written as

$$AF_N = \frac{A_{DC}F_N f_4}{\dfrac{f^4}{f_1 f_2 f_3} - \dfrac{f^2}{f_1} - \dfrac{f^2}{f_2} - \dfrac{f^2}{f_3} + j\left(f - \dfrac{f^3}{f_1 f_2} - \dfrac{f^3}{f_1 f_3} - \dfrac{f^3}{f_2 f_3}\right)} . \quad (7.17)$$

At $\underline{/AF_N} = -180°$, AF_N is real; hence, the imaginary part of the denominator is zero:

$$f - \frac{f^3}{f_1 f_2} - \frac{f^3}{f_1 f_3} - \frac{f^3}{f_2 f_3} = 0. \quad (7.18)$$

One solution of Equation 7.18 is $f = 0$, corresponding to the intersection of the line of AF_N with the positive real axis. If $f \neq 0$, Equation 7.18 can be divided by f, resulting in

$$f = \pm \sqrt{\frac{f_1 f_2 f_3}{f_1 + f_2 + f_3}} . \quad (7.19)$$

At these frequencies, one positive and one negative, AF_N is real with a value of

$$AF_N = \frac{-A_{DC}F_N f_4 (f_1 + f_2 + f_3)^2}{2 f_1 f_2 f_3 + f_1^2 f_2 + f_1^2 f_3 + f_2^2 f_1 + f_2^2 f_3 + f_3^2 f_1 + f_3^2 f_2} . \quad (7.20)$$

The feedback amplifier is stable if this value is more positive than -1, that is, if

$$A_{DC}F_N < \frac{2 f_1 f_2 f_3 + f_1^2 f_2 + f_1^2 f_3 + f_2^2 f_1 + f_2^2 f_3 + f_3^2 f_1 + f_3^2 f_2}{f_4 (f_1 + f_2 + f_3)^2} . \quad (7.21)$$

Example 7.3. A Type 702A operational amplifier is characterized by an amplification A of Equation 5.15 with $f_1 = 1$ MHz, $f_2 = 4$ MHz, $f_3 = 40$ MHz, and $A_{DC} = 4000$. The operational amplifier is used as a voltage follower, i.e., feedback return $F_N = 1$; hence, $A_{DC}F_N = 4000$. It was shown in Example 7.2 that under these conditions the feedback amplifier circuit is not stable. In order to obtain a stable system, a fourth lag network with $f_4 = 1$ kHz will be incorporated in AF_N. With these values of f_1, f_2, f_3, and f_4, the condition of stability given by

Equation 7.21 becomes $A_{DC}F_N < = 4454$. Thus, an $A_{DC}F_N = 4000$ results in a stable system.

Criteria of stability can likewise be computed for some other forms of amplification A. A number of these are summarized in Table 3 on page 140.

THE GENERAL CASE

It was seen that in the case of an amplification A consisting of lag networks, the analytical criterion of stability may result in long expressions. Furthermore, for six or more corner frequencies, general analytical solutions are not possible. The situation is similar if there is a fixed frequency-independent time delay, such as a length of terminated coaxial cable, included in A. In such cases it is necessary to revert to the basic Nyquist criterion of stability and to graphical evaluation, such as the Nyquist diagram or the Bode plots.

PROBLEMS

1. Derive Equations 7.6, 7.8, and 7.10.

2. Derive entries 3 and 4 in Table 3, page 140, from entry 5.

3. A Type 741 internally compensated operational amplifier can be characterized by an A of the form of Equation 5.15 with $f_1 = 10$ Hz, $f_2 = f_3 = 10$ MHz, and $A_{DC} = 200,000$. Show that the operational amplifier is stable without additional compensation when it is used as a voltage follower.

4. Derive Equations 7.17 and 7.21.

5. An operational amplifier is characterized by an A in the form of Equation 7.14 with $f_1 = 1$ MHz, $f_2 = 4$ MHz, $f_3 = 40$ MHz, $f_4 = 100$ MHz, and $A_{DC} = 4000$. Sketch the Bode plots of AF_N if $A_{DC}F_N = 40$. Utilize the Bode plots and sketch an approximate Nyquist diagram. Is the system stable?

6. An operational amplifier includes a frequency-independent time delay and is characterized by an $A = 1000\,e^{-jf/f_D}/(1 + jf/f_1)$. Plot the Bode plots of AF_N, assuming $F_N = 1$, and $f_D/f_1 = 1$, 100, and 10,000. Discuss stability conditions.

7. Derive entry 6 of Table 3, page 140, from entry 7.

8. Derive entry 8 of Table 3, page 141 from entry 9.

9. Sketch the Bode plots of AF_N of Example 7.3 by using piecewise linear approximation.

10. Sketch the gain plot of AF_N of Problem 9, but without using piecewise linear approximation.

11. Demonstrate that all stability considerations derived in this chapter for noninverting feedback amplifiers can be applied to inverting feedback amplifiers if $AF_I/(1 + F_I)$ is substituted in place of AF_N.

12. An approximate rule of stability states that a system is stable if the Bode plot of the gain of $|AF_N|$, or of $|AF_I/(1 + F_I)|$, crosses the 0-dB axis with a slope that is not steeper than -40 dB/decade. Show that this rule is exact if AF_N is of the form of Equation 7.4 with $f_1 \ll f_2 \leq f_3$. Show that the error resulting from the application of this rule in the maximum allowed $A_{DC}F_N$ is -35% if AF_N is in the form of Equation 7.4 with $f_1 = f_2 = f_3$. What is the error if $f_1 \ll f_2 = f_3$? Show that the rule may break down completely in a case such as the one discussed in Problem 6 above.

13. A noninverting feedback amplifier circuit is characterized by an $AF_N = A_{DC}F_N(1 + jf/10 \text{ MHz})^2/(1 + jf/100 \text{ kHz})^3$. Show that the system is stable if $A_{DC}F_N = 10^9$ and that it is unstable if $A_{DC}F_N = 10^3$. Sketch the Nyquist diagrams for both cases. Utilize the approximation of Entry 10 of Table 3 on page 141 to find the *minimum* allowed value of $A_{DC}F_N$ if a stable system is desired.

8

○○○○○○○○○○○○○○○○○○○○○○○○○○○○○○○○○○○○○○

Compensation Techniques

It was seen in Example 7.3 of the preceding chapter how an amplification A that would result in an unstable system could be modified by an additional lag network in order to arrive at a stable system. This modification is denoted *lag compensation*; other principal compensation methods are *modified lag compensation*, *lead compensation*, and *lead-lag compensation*.

LAG COMPENSATION

When an operational amplifier with an amplification A is utilized as a noninverting feedback amplifier, the resulting amplification is

$$M_N = \frac{A}{1 + AF_N}. \tag{8.1}$$

Depending on the characteristics of AF_N, the system may or may not be stable. If the operational amplifier without feedback is stable and it can be represented by a frequency-dependent amplification A, then it can be shown that the resulting amplification of the feedback amplifier circuit, M_N, can be always made stable by modifying (compensating) A by the addition of a lag network (Fig. 5.1), resulting in a compensated $A_{\mathrm{comp}}F_N$ of

$$A_{\mathrm{comp}}F_N = \frac{AF_N}{1 + \mathrm{j}f/f_0}. \tag{8.2}$$

Example 8.1. The amplification of an operational amplifier can be represented as $A = A_{DC}/(1 + jf/f_1)^4$ with $f_1 = 1$ MHz and $A_{DC} = 10,000$. The amplifier is used as a noninverting feedback amplifier with a feedback return of $F_N = 0.1$.

Without any compensation, at a frequency of f_1 the phase $\underline{/A} = -4 \times 45° = -180°$ and the absolute value $|AF_N| = 10,000 \times 0.1/4 = 250 > 1$. Thus, the system is not stable.

By modifying A by the addition of a lag network with a corner frequency of $f_0 = 100$ Hz (see Equation 8.2), the compensated $|A_{\text{comp}}F_N|$ becomes unity at approximately $f_0 A_{DC} F_N = 100$ Hz $\times 10,000 \times 0.1 = 100$ kHz. At this frequency $\underline{/AF_N} \approx -113°$, hence, the system is stable.

In the case of an operational amplifier with an amplification of

$$A = \frac{A_{DC}}{(1 + jf/f_1)^3}, \tag{8.3}$$

compensated by the addition of a lag network with a corner frequency of f_0, the resulting (compensated) $A_{\text{comp}}F_N$ is

$$A_{\text{comp}}F_N = \frac{A_{DC}F_N}{(1 + jf/f_1)^3(1 + jf/f_0)}. \tag{8.4}$$

It is seen from entry 7 of Table 3, page 141, that this results in a stable system if

$$A_{DC}F_N < 8\frac{(1 + f_1/f_0)^3}{(1 + 3f_1/f_0)^2}. \tag{8.5}$$

For a stable system with $A_{DC}F_N \gg 1$ it can be shown that $f_0 \ll f_1$ and that the condition of Equation 8.5 becomes

$$A_{DC}F_N < \frac{8}{9}\frac{f_1}{f_0}. \tag{8.6}$$

Hence, for a given feedback factor $A_{DC}F_N \gg 1$ and for a given f_1,

$$f_0 < \frac{8}{9}\frac{f_1}{A_{DC}F_N} \tag{8.7}$$

is the criterion of stability.

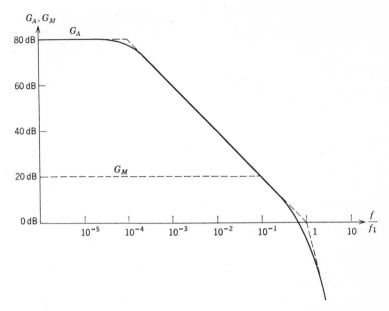

Figure 8.1. Bode plots $G_A \equiv 20$ dB $\log_{10} |A_{\text{comp}}|$ and $G_M \equiv 20$ dB $\log_{10} |M_N|$ for a feedback amplifier circuit with a lag compensated amplification of $A_{\text{comp}} = 10,000(1 + jf/f_1)^{-3}(1 + 10,000jf/f_1)^{-1}$, and with a feedback return of $F_N = 0.1$.

In principle, a small margin in the condition of Equation 8.7 should be adequate to provide a stable system. In reality, because of tolerances and variations, an f_0 much lower than the value given by Equation 8.7 should be chosen. It will be seen, however, that the lower the value of f_0, the lower the resulting bandwidth B; thus, compromise is often necessary.

Bode plots are shown in Fig. 8.1 for the case where $A_{DC} = 10,000$, $F_N = 0.1$, and $f_0 = f_1/10,000$. This choice of f_0 assures a safety margin of almost a factor of ten in $A_{DC}F_N$, which should be adequate to take care of expected variations in A_{DC}. It is seen that the bandwidth of the resulting feedback amplifier circuit is in the vicinity of $B \approx f_0 A_{DC} F_N = 0.1f_1 < \frac{8}{9}f_1$. Thus, the more conservative the choice of f_0, the smaller the bandwidth B. In the marginal case of $f_0 = \frac{8}{9}(f_1/A_{DC}F_N)$, the bandwidth is $B \approx \frac{8}{9}f_1$.

Example 8.2. An operational amplifier has an amplification of $A = A_{DC}/(1 + jf/f_1)^3$ with $A_{DC} = 10{,}000$ and $f_1 = 1$ MHz. It is used in a noninverting feedback amplifier circuit with a feedback return of $F_N = 0.01$; thus, $A_{DC}F_N = 10{,}000 \times 0.01 = 100$ and the system is not stable. In order to attain a stable system, lag compensation is introduced with a corner frequency of

$$f_0 < \frac{8}{9} \frac{f_1}{A_{DC}F_N} = \frac{8}{9} \frac{1\ \text{MHz}}{100} = 8.9\ \text{kHz}.$$

If an $f_0 = 8.9$ kHz were chosen (a marginal choice), the bandwidth of the feedback amplifier would be in the vicinity of

$$B \approx \tfrac{8}{9} f_1 = 0.89\ \text{MHz}.$$

If a conservative $f_0 = 0.89$ kHz is chosen, the bandwidth B of the feedback amplifier will be in the vicinity of

$$B \approx f_0 A_{DC} F_N = 0.89\ \text{kHz} \times 100 = 89\ \text{kHz}.$$

In the case of an amplifier with an amplification A consisting of three different lag networks, for $A_{DC}F_N \gg 1$ the required corner frequency f_0 of the lag compensation can be determined from entry 9 of Table 3, page 141.

Example 8.3. A Type 702A operational amplifier is characterized by an amplification A of Equation 5.15 with $f_1 = 1$ MHz, $f_2 = 4$ MHz, $f_3 = 40$ MHz, and $A_{DC} = 4000$. It is used as a noninverting feedback amplifier with a feedback return of $F_N = 0.1$; hence, the feedback factor $A_{DC}F_N = 4000 \times 0.1 = 400$. By substituting f_1, f_2, f_3, and $A_{DC}F_N$ into the criterion of stability, Entry 9 of Table 3, page 141, an $f_0 < 11$ kHz results. In a realistic design an f_0 of much less than 11 kHz would be used. If, however, an $f_0 = 11$ kHz were chosen, then the resulting Bode plots would be those shown in Fig. 8.2. It is seen that the magnitude of the compensated $A_{\text{comp}}F_N$ is unity at a frequency of $f \approx 2$ MHz and the phase φ is $-180°$ at the same frequency. Thus, the Nyquist diagram crosses the negative real axis *at the $-1 + j\,0$ point and the system is marginally stable*. It

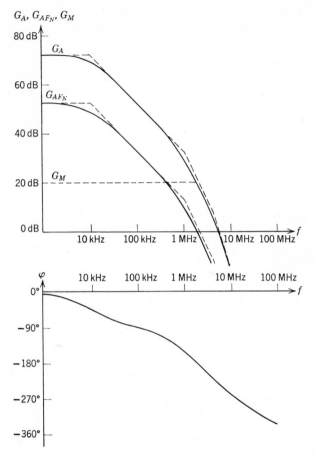

Figure 8.2. Bode plots $G_A \equiv 20 \text{ dB} \log_{10} |A_{\text{comp}}|$, $G_{AF_N} \equiv 20 \text{ dB} \log_{10} |A_{\text{comp}}F_N|$, $G_M \equiv 20 \text{ dB} \log_{10} |M_N|$, and $\varphi \equiv \underline{/A_{\text{comp}}} = \underline{/A_{\text{comp}}F_N}$ for Example 8.3 with a lag compensated $A_{\text{comp}} = 4000(1 + jf/1 \text{ MHz})^{-1}(1 + jf/4 \text{ MHz})^{-1} \times (1 + jf/40 \text{ MHz})^{-1}(1 + jf/11 \text{ kHz})^{-1}$, and with a feedback return of $F_N = 0.1$.

can be also seen from the Bode plot of $|M_N|$ that the bandwidth of the amplifier is in the vicinity of 2 MHz.

MODIFIED LAG COMPENSATION

The bandwidth of a feedback amplifier circuit can be increased somewhat by using a modified lag network of the form of Fig. 5.5 instead of a simple lag network. A modified lag network can be characterized by a transfer function in the form of Equation 5.17a, that is, by a

$$\frac{V_{\text{out}}(f)}{V_{\text{in}}(f)} = \frac{1 + jf/f_1}{1 + jf/f_0}, \qquad f_0 < f_1. \tag{8.8}$$

By choosing f_1 equal to the lowest corner frequency of A, some advantage can be gained over the simple lag compensation.

In the case of an operational amplifier with an amplification A of Equation 8.3,

$$A = \frac{A_{DC}}{(1 + jf/f_1)^3}, \tag{8.9}$$

and with the application of the modified lag network of Equation 8.8, the compensated $A_{\text{comp}}F_N$ becomes

$$A_{\text{comp}}F_N = \frac{A_{DC}F_N}{(1 + jf/f_1)^3}\frac{1 + jf/f_1}{1 + jf/f_0} = \frac{A_{DC}F_N}{(1 + jf/f_1)^2(1 + jf/f_0)}. \tag{8.10}$$

From entry 4 of Table 3, page 140, the criterion of stability is

$$A_{DC}F_N < 4 + 2\left(\frac{f_0}{f_1} + \frac{f_1}{f_0}\right). \tag{8.11}$$

If $f_0 \ll f_1$, this becomes

$$A_{DC}F_N < 2\frac{f_1}{f_0}, \tag{8.12}$$

and the bandwidth of the feedback amplifier circuit is in the vicinity of $f_0 A_{DC}F_N$. Thus, if the marginal case of $f_0 = 2f_1/A_{DC}F_N$ were chosen, then the 3-dB bandwidth would be approximately $B \approx 2f_1$, an improvement over the bandwidth of $B \approx \frac{8}{9}f_1$ of the simple lag compensation.

Example 8.4. An operational amplifier has an amplification of $A = A_{DC}/(1 + jf/f_1)^3$ with $A_{DC} = 10,000$ and $f_1 = 1$ MHz. The amplifier is used as a noninverting feedback amplifier with a feedback return of $F_N = 0.01$; thus, $A_{DC}F_N = 10,000 \times 0.01 = 100$ and the system is not stable. In order to attain a stable system, modified lag compensation of $(1 + jf/f_1)/(1 + jf/f_0)$ is introduced with

$$f_0 < \frac{2f_1}{A_{DC}F_N} = \frac{2 \times 1 \text{ MHz}}{100} = 20 \text{ kHz}.$$

If an $f_0 = 20$ kHz were chosen (a marginal choice), the maximum bandwidth of the feedback amplifier circuit would be in the vicinity of $B \approx 2f_1 = 2 \times 1$ MHz $= 2$ MHz, an improvement over the 0.89-MHz bandwidth of the lag compensated amplifier (see Example 8.2).

In the case when amplification A of the operational amplifier is in the form of Equation 5.15, that is,

$$A = \frac{A_{DC}}{(1 + jf/f_1)(1 + jf/f_2)(1 + jf/f_3)} \tag{8.13}$$

with $f_1 \neq f_2 \neq f_3$, and a modified lag compensation of the form $(1 + jf/f_1)/(1 + jf/f_0)$ is introduced. This will result in a compensated $A_{comp}F_N$ of

$$A_{comp}F_N = \frac{A_{DC}F_N}{(1 + jf/f_1)(1 + jf/f_2)(1 + jf/f_3)} \frac{1 + jf/f_1}{1 + jf/f_0}$$

$$= \frac{A_{DC}F_N}{(1 + jf/f_0)(1 + jf/f_2)(1 + jf/f_3)}. \tag{8.14}$$

If $f_0 \ll f_2$ and $f_0 \ll f_3$, then the criterion of stability, from entry 5 of Table 3, page 140, becomes

$$A_{DC}F_N < \frac{f_2 + f_3}{f_0} + \frac{f_3}{f_2} + \frac{f_2}{f_3}, \tag{8.15}$$

that is,

$$f_0 < \frac{f_2 + f_3}{A_{DC}F_N - f_3/f_2 - f_2/f_3}. \tag{8.16}$$

Example 8.5. A Type 702A operational amplifier is characterized by an amplification A of Equation (8.13) with $f_1 = 1$ MHz, $f_2 = 4$ MHz, $f_3 = 40$ MHz, and $A_{DC} = 4000$. The amplifier is used as a noninverting feedback

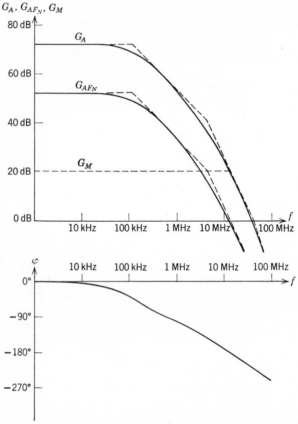

Figure 8.3. Bode plots $G_A \equiv 20$ dB $\log_{10}|A_{\text{comp}}|$, $G_{AF_N} \equiv 20$ dB $\log_{10}|A_{\text{comp}}F_N|$, $G_M \equiv 20$ dB $\log_{10}|M_N|$, and $\varphi \equiv \underline{/A_{\text{comp}}} = \underline{/A_{\text{comp}}F_N}$ for Example 8.5 with a modified lag compensated

$$A_{\text{comp}} = 4000(1 + \mathrm{j}f/4 \text{ MHz})^{-1}(1 + \mathrm{j}f/40 \text{ MHz})^{-1} \times (1 + \mathrm{j}f/110 \text{ kHz})^{-1},$$

and with a feedback return of $F_N = 0.1$.

amplifier with a feedback return of $F_N = 0.1$; hence, $A_{DC}F_N = 4000 \times 0.1 = 400$ and the system is not stable. In order to attain a stable system, a modified lag network of the form $(1 + jf/f_1)/(1 + jf/f_0)$ is introduced. The criterion of stability, Equation 8.16, thus becomes

$$f_0 < \frac{f_2 + f_3}{A_{DC}F_N - f_3/f_2 - f_2/f_3}$$

$$= \frac{4 \text{ MHz} + 40 \text{ MHz}}{400 - 40 \text{ MHz}/4 \text{ MHz} - 4 \text{ MHz}/40 \text{ MHz}}$$

$$\approx 110 \text{ kHz}.$$

In a realistic design, an f_0 of much less than 110 kHz would be chosen; for purposes of comparison, however, an $f_0 = 110$ kHz will be assumed here. The Bode plots thus resulting are shown in Fig. 8.3. The magnitude of the compensated $A_{comp}F_N$ is unity at a frequency of $f \approx 12$ MHz, and the phase φ is $-180°$ at the same frequency. Thus, the Nyquist diagram crosses the negative real axis *at* the $-1 + j\,0$ point, and the system is marginally stable. It can also be seen from the Bode plot of $|M_N|$ that the bandwidth of the feedback amplifier circuit is now in the vicinity of 12 MHz, a significant improvement over the 2-MHz bandwidth of the lag compensated amplifier (see Example 8.3.).

LEAD COMPENSATION

Lead compensation consists of the inclusion of a lead network of the form of Fig. 5.7 modifying AF_N. The transfer function of such a network can be written, by utilizing Equation 5.23a, as

$$\frac{V_{out}(f)}{V_{in}(f)} = \frac{f_1}{f_2}\frac{1 + jf/f_1}{1 + jf/f_2}, \qquad f_1 < f_2. \qquad (8.17)$$

By choosing f_1 to equal the lowest corner frequency of A, in some cases an advantage may be gained over lag compensation and modified lag compensation techniques.

In the simple case of an operational amplifier with an amplification given by A of Equation 8.3,

$$A = \frac{A_{DC}}{(1 + jf/f_1)^3} \qquad (8.18)$$

and with the application of the lead network of Equation 8.17, the compensated $A_{comp}F_N$ can be written as

$$A_{comp}F_N = \frac{A_{DC}F_N}{(1 + jf/f_1)^3} \frac{1 + jf/f_1}{1 + jf/f_2} = \frac{A_{DC}F_N}{(1 + jf/f_1)^2(1 + jf/f_2)}, \qquad (8.19)$$

where $f_2 > f_1$.* Also, in practical circuits, the limitation $F_N \leq f_1/f_2$ has to be observed when the lead compensation is external to the operational amplifier.

If $f_2 \gg f_1$, the criterion of stability reduces to

$$A_{DC}F_N < 2\frac{f_2}{f_1}, \qquad (8.20a)$$

that is, to

$$f_2 > A_{DC}F_N \frac{f_1}{2}. \qquad (8.20b)$$

It can be shown by the inspection of the Bode plots, that for $f_2 \geq B$ the bandwidth of the feedback amplifier is in the vicinity of

$$B \approx f_1\sqrt{A_{DC}F_N}. \qquad (8.21)$$

Example 8.6. An operational amplifier has an amplification of $A = A_{DC}/(1 + jf/f_1)^3$ with $A_{DC} = 10,000$ and $f_1 = 1$ MHz. It is used as a noninverting feedback amplifier with a feedback return of $F_N = 0.01$; thus, the feedback factor is $A_{DC}F_N = 10,000 \times 0.01 = 100$ and the system is not stable. A lead compensation of the form $(1 + jf/f_1)/(1 + jf/f_2)$ is introduced with $f_1 = 1$ MHz. From Equation 8.20b, the criterion of stability requires an

$$f_2 > A_{DC}F_N \frac{f_1}{2} = 100\frac{1 \text{ MHz}}{2} = 50 \text{ MHz}.$$

---

* Here, and in what follows, the constant factor f_1/f_2 of Equation 8.17 is included in F_N.

This criterion can be satisfied, for example, by a choice of $f_2 = 100$ MHz. The resulting bandwidth of the feedback amplifier circuit is then in the vicinity of

$$B \approx f_1 \sqrt{A_{DC} F_N} = 1 \text{ MHz} \sqrt{100} = 10 \text{ MHz}.$$

This is a significant improvement over the $B = 0.89$ MHz bandwidth of the lag compensated amplifier (Example 8.2), and over the 2-MHz bandwidth of the modified lag compensated amplifier (Example 8.4).

It is seen that the use of lead compensation results in an improvement of the bandwidth by a factor of $0.5\sqrt{A_{DC} F_N}$ over the modified lag compensation; this can be significant when $A_{DC} F_N$ is large. Unfortunately, the stability of the feedback amplifier circuit with lead compensation is critically dependent upon the exact cancellation of the two factors of $(1 + jf/f_1)$. If the capacitance of the lead network is adjustable, this difficulty is easily overcome by adjusting this capacitance to attain an exact cancellation; if, however, adjustment of the capacitance is not feasible, the circuit may become impractical. Also, the lead compensation may be seriously affected by stray capacitances.

Example 8.7. The feedback amplifier circuit shown in Fig. 8.4 uses $R_I = 200\,\Omega$, $R_F = 19,800\,\Omega$ and an operational amplifier with an amplification of $A = A_{DC}/(1 + jf/f_1)^3$, where $A_{DC} = 10,000$ and $f_1 = 1$ MHz. Feedback return

$$F_N = \frac{R_I}{R_I + R_F} = \frac{200\,\Omega}{200\,\Omega + 19,800\,\Omega} = 0.01;$$

hence, $A_{DC} F_N = 10,000 \times 0.01 = 100$. This is larger than the 8 of entry 3 in Table 3, page 140, thus, the feedback amplifier circuit without compensation (i.e., with $C = 0$) is unstable.

By using the lead compensation of Equation 5.23a with $f_1 = 1/2\pi C R_F = 1$ MHz (see Equation 5.23b), the compensated $A_{\text{comp}} F_N$ can be written as

$$A_{\text{comp}} F_N = \frac{A_{DC} F_N}{(1 + jf/f_1)^3} \frac{1 + jf/f_1}{1 + jf/f_2} = \frac{A_{DC} F_N}{(1 + jf/f_1)^2 (1 + jf/f_2)}.$$

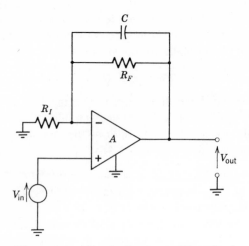

Figure 8.4. The feedback amplifier circuit of Example 8.7.

The value of f_2, by utilizing Equation 5.23c, is

$$f_2 = \frac{1}{2\pi C \dfrac{R_F R_I}{R_I + R_F}}$$

$$= \frac{1}{2\pi C R_F F_N} = \frac{f_1}{F_N} = \frac{1 \text{ MHz}}{0.01} = 100 \text{ MHz}.$$

The value of capacitance C is therefore,

$$C = \frac{1}{2\pi R_F f_1} = \frac{1}{2\pi \times 19{,}800 \times 10^6} = 8 \text{ pF}.$$

From entry 4 of Table 3, page 140, the feedback amplifier circuit will be stable if

$$A_{DC} F_N < 4 + 2 \left(\frac{f_2}{f_1} + \frac{f_1}{f_2} \right)$$

$$= 4 + 2 \left(\frac{100 \text{ MHz}}{1 \text{ MHz}} + \frac{1 \text{ MHz}}{100 \text{ MHz}} \right)$$

$$= 204.$$

Thus, since $A_{DC}F_N = 100$, the circuit with lead compensation seems to be stable with a comfortable margin.

In reality, however, there will be stray capacitances in parallel with R_I; assume these total a $C_P = 10$ pF. Therefore, the voltage divider representation of Fig. 5.9 and Equation 5.28 must be used. By utilizing Equation 5.28, the transfer function of the "lead network" becomes

$$\frac{R_I}{R_I + R_F} \frac{1 + jf/1 \text{ MHz}}{1 + jf/45 \text{ MHz}}.$$

Thus, in the expression of the compensated $A_{\text{comp}}F_N$ above, $f_2 = 45$ MHz, while f_1 remains 1 MHz. Now, from entry 4 of Table 3, page 140, the criterion of stability is

$$A_{DC}F_N < 4 + 2\left(\frac{f_2}{f_1} + \frac{f_1}{f_2}\right)$$

$$= 4 + 2\left(\frac{45 \text{ MHz}}{1 \text{ MHz}} + \frac{1 \text{ MHz}}{45 \text{ MHz}}\right)$$

$$= 94.$$

Therefore, since $A_{DC}F_N = 100$, the resulting feedback amplifier circuit is not stable.

The effects of C_P can be reduced if the values of resistances R_I and R_F can be made smaller. If $R_I = 20 \ \Omega$, $R_F = 1980 \ \Omega$, and $C = 80$ pF are used, then it can be shown that the transfer function of the "lead network" is

$$\frac{R_I}{R_I + R_F} \frac{1 + jf/1 \text{ MHz}}{1 + jf/90 \text{ MHz}}$$

and the criterion of stability, from entry 4 of Table 3 on page 140, becomes

$$A_{DC}F_N < 4 + 2\left(\frac{90 \text{ MHz}}{1 \text{ MHz}} + \frac{1 \text{ MHz}}{90 \text{ MHz}}\right) \approx 184.$$

Thus, in this case, since $A_{DC}F_N = 100$, the system is stable.

LEAD-LAG COMPENSATION

Some advantages of the lead and lag compensation techniques may be combined by the use of lead-lag compensation. Lead-lag compensation, as the name implies, consists of the inclusion of a lead and a lag network modifying AF_N or AF_I. The transfer function of a lead-lag compensation may be of the form

$$\text{constant} \times \frac{1 + jf/f_1}{1 + jf/f_2} \frac{1}{1 + jf/f_0}, \qquad f_0 \ll f_1 < f_2;$$

the first factor represents a lead compensation, the second one a lag compensation.

In the simple case of an operational amplifier with an amplification A of Equation 8.3, that is, with an

$$A = \frac{A_{DC}}{(1 + jf/f_1)^3}, \tag{8.22}$$

and with the application of a lead-lag network, the compensated $A_{\text{comp}}F_N$ becomes

$$\begin{aligned}
A_{\text{comp}}F_N &= \frac{A_{DC}F_N}{(1 + jf/f_1)^3} \frac{1 + jf/f_1}{1 + jf/f_2} \frac{1}{1 + jf/f_0} \\
&= \frac{A_{DC}F_N}{(1 + jf/f_1)^2(1 + jf/f_2)(1 + jf/f_0)}. \tag{8.23}
\end{aligned}$$

The criterion of stability for $f_0 \ll f_1 < f_2$, from entry 8 of Table 3, page 141, is

$$A_{DC}F_N < \frac{2f_1}{f_0} \frac{(1 + f_1/f_2)^2}{(1 + 2f_1/f_2)^2}. \tag{8.24}$$

If, furthermore,

$$f_1 \ll \frac{f_2}{2},$$

Equation 8.24 becomes

$$A_{DC}F_N < \frac{2f_1}{f_0}, \tag{8.25}$$

providing no advantage over the case of the modified lag compensation. For this reason, lead-lag compensation in this form is seldom used.

Of more utility is the lead-lag compensation that may be designated as lead-modified lag compensation, that is, the inclusion of a network, or networks, with a resulting transfer function in the form

$$\text{constant} \times \frac{1 + jf/f_1}{1 + jf/f_2} \frac{1 + jf/f_3}{1 + jf/f_0}.$$

The first factor represents a lead compensation, the second one a modified lag compensation. In practice, the two numerators are sometimes switched in the circuit, with no effect on the resulting transfer function.

In the simple case of an operational amplifier with the amplification of Equation 8.3, that is, with an

$$A = \frac{A_{DC}}{(1 + jf/f_1)^3}, \tag{8.26}$$

and by the application of a lead-modified lag network with a transfer function of the form

$$\text{constant} \times \frac{1 + jf/f_1}{1 + jf/f_2} \frac{1 + jf/f_1}{1 + jf/f_0}, \qquad f_0 \ll f_1 < f_2,$$

the compensated $A_{\text{comp}}F_N$ becomes

$$
\begin{aligned}
A_{\text{comp}}F_N &= \frac{A_{DC}F_N}{(1 + jf/f_1)^3} \frac{1 + jf/f_1}{1 + jf/f_2} \frac{1 + jf/f_1}{1 + jf/f_0} \\
&= \frac{A_{DC}F_N}{(1 + jf/f_1)(1 + jf/f_2)(1 + jf/f_0)}.
\end{aligned}
\tag{8.27}
$$

If $f_2 \gg f_1$ and $f_2 \gg f_0$, then the criterion of stability from entry 5 of Table 3, page 140, becomes

$$A_{DC}F_N < \frac{f_2}{f_1} + \frac{f_2}{f_0}. \tag{8.28}$$

Example 8.8. An operational amplifier has an amplification of $A = A_{DC}/(1 + jf/f_1)^3$ with $A_{DC} = 10{,}000$ and $f_1 = 1$ MHz. The amplifier is used as a noninverting feedback amplifier with a feedback return of $F_N = 0.01$; hence, $A_{DC}F_N = 10{,}000 \times 0.01 = 100$. If lead-modified lag compensation is used with an $f_2 = 100$ MHz and with a (conservative) $f_0 = 0.1$ MHz, the criterion of stability, from Equation 8.28, becomes $A_{DC}F_N < 100$ MHz/1 MHz + 100 MHz/0.1 MHz = 1100. Thus, since $A_{DC}F_N = 100$, the system is stable; also the 3-dB bandwidth is in the vicinity of $B \approx f_0 A_{DC}F_N = 0.1$ MHz $\times 100 = 10$ MHz. This is a significant improvement over the 2-MHz bandwidth with no margin of the modified lag compensated amplifier (see Example 8.4). Because of the better stability margin, it is also an improvement over the lead compensated amplifier (see Example 8.6) that has a bandwidth in the vicinity of 10 MHz.

The lead-modified lag compensation is particularly advantageous when amplification A has different corner frequencies, that is, when it is in the form of Equation 5.15:

$$A = \frac{A_{DC}}{(1 + jf/f_1)(1 + jf/f_2)(1 + jf/f_3)} \qquad (8.29)$$

with $f_1 < f_2 < f_3$. Introducing a lead-modified lag compensation with a transfer function of the form

$$\text{constant} \times \frac{1 + jf/f_2}{1 + jf/f_3} \frac{1 + jf/f_1}{1 + jf/f_0},$$

the compensated $A_{\text{comp}}F_N$ becomes

$$A_{\text{comp}}F_N = \frac{A_{DC}F_N}{(1 + jf/f_1)(1 + jf/f_2)(1 + jf/f_3)} \frac{1 + jf/f_2}{1 + jf/f_3} \frac{1 + jf/f_1}{1 + jf/f_0}$$

$$= \frac{A_{DC}F_N}{(1 + jf/f_3)^2(1 + jf/f_0)}. \qquad (8.30)$$

Example 8.9. A Type 702A operational amplifier is characterized by an amplification A of Equation 8.29 with $f_1 = 1$ MHz, $f_2 = 4$ MHz, $f_3 = 40$ MHz, and $A_{DC} = 4000$. It is used as a noninverting feedback amplifier with a feedback return of $F_N = 0.1$; hence, $A_{DC}F_N = 4000 \times 0.1 = 400$ and the system is not stable. In order to attain a stable system, a lead-modified lag network of

$$\frac{1 + jf/f_2}{1 + jf/f_3} \frac{1 + jf/f_1}{1 + jf/f_0}$$

with $f_0 < f_1$ is utilized. Thus, $A_{\text{comp}}F_N$ becomes

$$A_{\text{comp}}F_N = \frac{A_{DC}F_N}{(1 + jf/f_1)(1 + jf/f_2)(1 + jf/f_3)}\frac{1 + jf/f_2}{1 + jf/f_3}\frac{1 + jf/f_1}{1 + jf/f_0}$$

$$= \frac{A_{DC}F_N}{(1 + jf/f_3)^2(1 + jf/f_0)}.$$

For $f_0 \ll f_3$, by utilizing entry 4 of Table 3, page 140, the criterion of stability becomes

$$A_{DC}F_N < 2\frac{f_3}{f_0},$$

that is,

$$f_0 < \frac{2f_3}{A_{DC}F_N} = \frac{2 \times 40\,\text{MHz}}{400} = 200\,\text{kHz}.$$

It can be shown that if a marginal $f_0 = 200\,\text{kHz}$ were chosen, then the bandwidth of the feedback amplifier would be in the vicinity of $B = 50\,\text{MHz}$. This represents a significant improvement over the 12-MHz bandwidth of the modified lag compensated amplifier (see Example 8.5).

COMPENSATION CIRCUITS

The compensation techniques discussed above modify the frequency characteristics of the operational amplifier to achieve a desired margin of stability and a suitable bandwidth. It was seen that the resulting bandwidth in the case of the simple lag compensation could be increased by the utilization of modified lag compensation, and in some cases by the use of lead compensation or lead-lag compensation.

The simplest of the compensation techniques, the lag compensation, is usually the easiest to implement. In some cases, such as those shown in Fig. 8.5 and in Fig. 8.6 in this chapter's Problems, the low corner frequency f_0 of the lag compensation is implemented— by necessity of the configuration—external to the operational amplifier. In many cases, however, corner frequency f_0 of the lag compensation is implemented by connecting an external capacitance between ground and a point within the operational amplifier such as the collector of a transistor amplifier stage.

Many operational amplifiers include a compensating lag network: these are the *internally compensated operational amplifiers*. In these amplifiers, the value of corner frequency f_0 is chosen such that the operational amplifier is stable when used as a voltage follower. However, corner frequency f_0 is now excessively low when the amplifier is used in other than voltage-follower applications, hence, the resulting bandwidth is inferior to that obtainable with external compensation. In applications where bandwidth requirements are not stringent this may not be a significant limitation and the use of internally compensated operational amplifiers should be considered.

When the bandwidth obtained by the use of lag compensation is inadequate, other compensation methods may be applied. Modified lag compensation can be implemented by using a series resistance-capacitance network instead of the capacitance of the simple lag network. Lead compensation is usually applied external to the operational amplifier, such as shown in Fig. 8.4. When lead-lag compensation is used, it is usually implemented by a combination of a modified lag network and a lead network, with one of them located external to the operational amplifier.

PROBLEMS

1. Demonstrate that a feedback amplifier circuit can be always made stable by the inclusion of a suitable lag network into A.

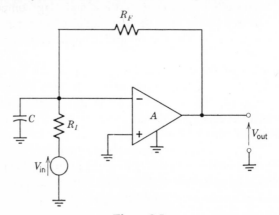

Figure 8.5.

2. Determine the minimum value of capacitance C in Fig. 8.5, if $A = 1000/(1 + jf/1\text{ MHz})^3$, $R_I = 100\ \Omega$, $R_F = 1000\ \Omega$, and if a stable system is desired.

3. The circuit of Fig. 8.6 utilizes a Type 702A operational amplifier with an amplification of

$$A = \frac{4000}{(1 + jf/1\text{ MHz})(1 + jf/4\text{ MHz})(1 + jf/40\text{ MHz})}.$$

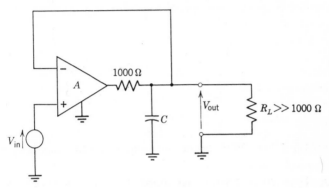

Figure 8.6.

Determine the minimum value of capacitance C if a stable system is desired.

4. The circuit of Fig. 8.7 consists of an operational amplifier with an amplification of $A = 1000/(1 + jf/f_1)$ and a delay line of length L terminated by its characteristic impedance R_0. The transfer function of the delay line can be represented as e^{-jf/f_0}, where $f_0 = 1/2\pi\tau L$ with $\tau \approx 1.5$ ns/foot (≈ 0.05 ns/cm). Determine the maximum length of the delay line if $f_1 = 1$ MHz and if a stable system is desired.

5. The circuit of Fig. 8.8 incorporates a voltage follower with an inductance L in series with the output of the operational amplifier. Determine the allowed ranges of values for the inductance such that the system is stable if the operational amplifier is an internally compensated Type 741 with an amplification of

$$A = \frac{200{,}000}{(1 + jf/10\text{ Hz})(1 + jf/10\text{ MHz})^2}.$$

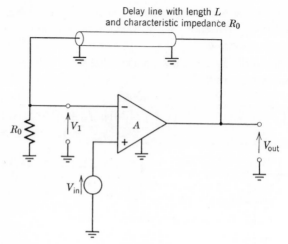

Figure 8.7.

Repeat for a Type 702A operational amplifier with amplification A as given in Problem 3.

6. A Type 702A operational amplifier is characterized by an amplification A of Equation 5.15 with $f_1 = 1$ MHz, $f_2 = 4$ MHz, $f_3 = 40$ MHz, and with a nominal dc amplification of $A_{DC} = 4000$. The operational amplifier is used as a voltage follower. Design a lag compensation, a modified lag compensation, and a lead-modified lag compensation such that the system is stable if $A_{DC} = 8000$,

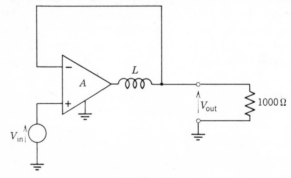

Figure 8.8.

but not stable if $A_{DC} > 8000$. Estimate the resulting bandwidths of $|M_N|$ for the various compensations assuming $A_{DC} = 4000$.

7. A Type 702A operational amplifier with an amplification A as given in Example 8.5 is utilized in the feedback amplifier circuit shown in Fig. 8.4 with $R_I = 10\,\Omega$ and $R_F = 1000\,\Omega$. Is the system stable if the value of capacitance C is chosen such that the $(1 + jf/1\,\text{MHz})$ factor in A is cancelled? Is the system stable if the value of capacitance C is chosen such that the $(1 + jf/4\,\text{MHz})$ factor in A is cancelled?

8. An operational amplifier is characterized by an amplification of

$$A = \frac{A_{DC}}{(1 + jf/1\,\text{MHz})(1 + jf/2\,\text{MHz})(1 + jf/5\,\text{MHz})(1 + jf/10\,\text{MHz})}.$$

It is used as a noninverting feedback amplifier with a feedback return of $F_N = 0.1$. Sketch the Bode plots and determine the maximum value of A_{DC} if a stable system is desired.

9. How do compensation considerations of an inverting feedback amplifier circuit differ from those of a noninverting feedback amplifier circuit?

10. The operational amplifier in the feedback amplifier circuit of Fig. 8.9 is characterized by an amplification of

$$A = \frac{500}{(1 + jf/1\,\text{MHz})(1 + jf/40\,\text{MHz})^2}.$$

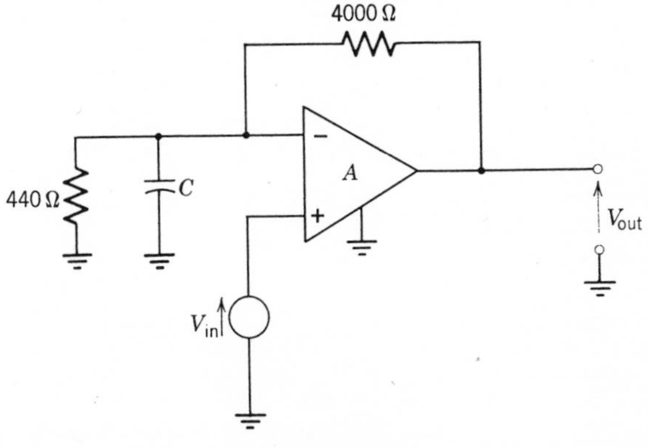

Figure 8.9.

Show that the feedback amplifier is stable if $C = 0$ or if $C = 5000$ pF and that it is unstable if $C = 10$ pF.

11. The feedback amplifier circuit shown in Fig. 8.4 uses $R_I = 100\ \Omega$, $R_F = 9900\ \Omega$, and an operational amplifier with an amplification of $A = A_{DC}/(1 + jf/f_1)^3$ where $A_{DC} = 1000$ and $f_1 = 1$ MHz. Show that the resulting feedback amplifier is not stable if $C = 0$. Choose C such that a factor of $(1 + jf/f_1)$ is cancelled in AF_N and show that the resulting feedback amplifier circuit is stable even if there is a 10-pF capacitance added in parallel with resistor R_I.

12. Sketch the Bode plot of $|M_N|$ for the circuit of Fig. 8.4 by using piecewise linear approximation. Assume $A = 10{,}000/(1 + jf/1\ \text{MHz})^3$, $R_I = 200\ \Omega$, $R_F = 19800\ \Omega$, and $C = 8$ pF.

13. Sketch the Bode plots of the phases $\underline{/A_{\text{comp}}}$ and $\underline{/M_N}$ and of the magnitude $|A_{\text{comp}}F_N|$ for the feedback amplifier circuit described by Fig. 8.1.

14. Sketch the Bode plots of the gains for Example 8.9 by using piecewise linear approximation.

15. Identify the source of the limitation $F_N \leq f_1/f_2$, following Equation 8.19. (See also Fig. 8.4.)

16. An alternate representation of the Bode plots, that in many cases is less time consuming to plot, consists of one gain plot and separate vertical scales for G_A and G_{AF_N}. Sketch the Bode plots of Fig. 8.2 and Fig. 8.3 using this representation.

9

OOOOOOOOOOOOOOOOOOOOOOOOOOOOOOOOOO

Linear Attributes of Real Operational Amplifiers

In the preceding, ideal operational amplifiers were assumed in accordance with Equations 2.1 and 2.2. In the case of a real operational amplifier, however, these equations can be considered only as approximations. In this chapter, and in the subsequent one, departures of real operational amplifiers from Equations 2.1 and 2.2 will be discussed.

COMMON MODE AMPLIFICATION AND COMMON MODE REJECTION

Consider the circuit of Fig. 9.1. If the amplifier is ideal, the output voltage is given by

$$V_{\text{out}} = AV_d \qquad (9.1)$$

and voltage V_c has no effect whatsoever on the output voltage. In reality, a small fraction of V_c finds its way to the output, that is,

$$V_{\text{out}} = AV_d + A_{CM}V_c, \qquad (9.2)$$

where A_{CM} is designated as *common mode amplification*. In the case of a real operational amplifier, $A_{CM} \neq 0$, but, usually, its magnitude is much smaller than that of A:

$$|A_{CM}| \ll |A|. \qquad (9.3)$$

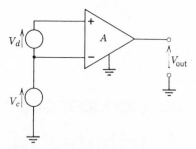

Figure 9.1. Circuit for determining the common mode rejection of an operational amplifier.

The output voltage of the amplifier, V_{out}, can be approximated as

$$V_{out} = D_d V_d + D_c V_c + K, \qquad (9.4a)$$

where D_d, D_c, and K are constants (in general, D_d, and D_c are the partial derivatives of V_{out} with respect to V_d and V_c, respectively). With the foregoing, the *common mode rejection ratio* (*CMRR*) is defined as

$$CMRR \equiv -20 \text{ dB} \log_{10} |D_c/D_d| = 20 \text{ dB} \log_{10} |D_d/D_c|. \qquad (9.4b)$$

It can be seen that for an ideal operational amplifier ($D_c = 0$), the common mode rejection ratio is $CMRR = \infty$. For the circuit of Fig. 9.1, $D_c = A_{CM}$ and $D_d = A$; hence, Equation 9.4b becomes

$$CMRR = -20 \text{ dB} \log_{10} |A_{CM}/A| = 20 \text{ dB} \log_{10} |A/A_{CM}|. \qquad (9.5)$$

In many cases, it is of interest to specify *CMRR* as function of frequency. In general, the common mode rejection ratio is best (*CMRR* the largest in magnitude) at dc, that is, at zero frequency, where

$$CMRR(f = 0) = 20 \text{ dB} \log_{10} |A(f = 0)/A_{CM}(f = 0)|$$
$$= 20 \text{ dB} \log_{10} |A_{DC}/A_{CM}(f = 0)|, \qquad (9.6)$$

designated as *dc common mode rejection ratio*, or in many instances simply as common mode rejection ratio.

Example 9.1. An operational amplifier is characterized by an amplification at zero frequency of $A_{DC} = 4000$ and by a common mode amplification at zero frequency of $|A_{CM}| = 0.04$. Thus, at zero frequency $|A_{CM}/A| = 0.04/4000 = 10^{-5}$, and the dc common mode rejection ratio becomes

$$CMRR(f = 0) = -20 \text{ dB log}_{10} (10^{-5})$$

$$= 20 \text{ dB log}_{10} (10^5)$$

$$= 100 \text{ dB}.$$

For many operational amplifiers, the common mode rejection ratio as a function of frequency can be approximated as

$$CMRR(f) \approx 20 \text{ dB log}_{10} \left| \frac{A_{DC}}{(1 + jf/f_{CM})A_{CM}(f = 0)} \right|$$

$$= 20 \text{ dB log}_{10} \left| \frac{A_{DC}}{A_{CM}(f = 0)} \right| - 20 \text{ dB log}_{10} |1 + jf/f_{CM}|$$

$$= CMRR(f = 0) - 20 \text{ dB log}_{10} \sqrt{1 + (f/f_{CM})^2}, \quad (9.7)$$

where f_{CM} is the corner frequency of the common mode rejection ratio.

Example 9.2. A Type 702A operational amplifier has a dc common mode rejection ratio of 95 dB. When driven from a zero-impedance source, as in Fig. 9.1, the corner frequency of the common mode rejection ratio is $f_{CM} = 0.5$ MHz. By utilizing Equation 9.7, the common mode rejection ratio as function of frequency becomes

$$CMRR(f) = CMRR(f = 0) - 20 \text{ dB log}_{10} \sqrt{1 + (f/f_{CM})^2}$$

$$= 95 \text{ dB} - 20 \text{ dB log}_{10} \sqrt{1 + (f/0.5 \text{ MHz})^2}.$$

Thus, for example, at a frequency of $f = 2$ MHz,

$$CMRR(f = 2 \text{ MHz})$$

$$= 95 \text{ dB} - 20 \text{ dB log}_{10} \sqrt{1 + (2 \text{ MHz}/0.5 \text{ MHz})^2}$$

$$= 95 \text{ dB} - 12.3 \text{ dB} = 82.7 \text{ dB}.$$

It is of importance to determine the common mode rejection properties of feedback amplifier circuits, in particular those of the differential feedback amplifier circuit with equalized amplifications (Fig. 3.4). If the operational amplifier if ideal, that is, if $A_{CM} = 0$, then the circuit, shown again in Fig. 9.2a, has an output voltage (see Equation 3.26):

$$V_{\text{out}} = \frac{A}{1 + \dfrac{R_S}{R_P} + A\dfrac{R_I}{R_F}\left(1 + \dfrac{R_S}{R_P}\right)\Big/\left(1 + \dfrac{R_I}{R_F}\right)} (V_d + V_c)$$

$$- \frac{A}{1 + \dfrac{R_I}{R_F} + A\dfrac{R_I}{R_F}} V_c. \quad (9.8)$$

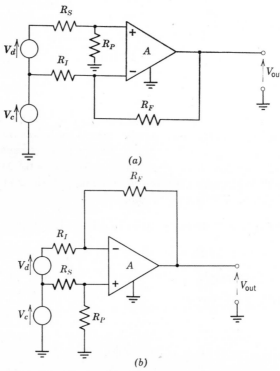

(a)

(b)

Figure 9.2. Two equivalent circuits for determining the common mode rejection of a differential amplifier circuit with feedback.

It can be shown that if $A_{CM} \neq 0$, then the output voltage in Fig. 9.2a is given by

$$V_{\text{out}} = \frac{A}{1 + \dfrac{R_S}{R_P} + A\dfrac{R_I}{R_F}\left(1 + \dfrac{R_S}{R_P}\right)\Big/\left(1 + \dfrac{R_I}{R_F}\right)}(V_d + V_c)$$

$$- \frac{A}{1 + \dfrac{R_I}{R_F} + A\dfrac{R_I}{R_F}}V_c + \frac{A_{CM}}{1 + \dfrac{R_I}{R_F} + A\dfrac{R_I}{R_F}}V_c. \quad (9.9)$$

Equation 9.9 will be now evaluated for several cases.

If $R_S/R_P = R_I/R_F$ and $A_{CM} = 0$, Equation 9.9 becomes

$$V_{\text{out}} = \frac{A}{1 + \dfrac{R_I}{R_F} + A\dfrac{R_I}{R_F}}V_d, \quad (9.10)$$

and the common mode rejection ratio of the circuit, $CMRR = \infty$.

If $R_S/R_P = R_I/R_F$, but $A_{CM} \neq 0$, then Equation 9.9 becomes

$$V_{\text{out}} = \frac{AV_d + A_{CM}V_c}{1 + \dfrac{R_I}{R_F} + A\dfrac{R_I}{R_F}}, \quad (9.11)$$

and the common mode rejection ratio of the circuit, by utilizing Equation 9.4, is

$$CMRR = -20 \text{ dB} \log_{10}|A_{CM}/A| = 20 \text{ dB} \log_{10}|A/A_{CM}|, \quad (9.12)$$

the same as that of an amplifier without feedback. Thus, if the resistors are perfectly balanced, the feedback has no effect on the common mode rejection ratio.

If $A_{CM} = 0$ but $R_S/R_P \neq R_I/R_F$, it can be shown by utilizing Equations 9.4 and 9.8 that the common mode rejection ratio of the circuit,

$$CMRR = -20 \text{ dB} \log_{10}\left|\left(1 - \frac{R_S}{R_P}\frac{R_F}{R_I}\right)\Big/\left(1 + \frac{R_F}{R_I}\right)\right|$$

$$= 20 \text{ dB} \log_{10}\left|\left(1 + \frac{R_F}{R_I}\right)\Big/\left(1 - \frac{R_S}{R_P}\frac{R_F}{R_I}\right)\right|. \quad (9.13)$$

In the case when $A_{CM} \neq 0$, $R_S/R_P \neq R_I/R_F$, but

$$|(R_S/R_P - R_I/R_F)/(R_S/R_P + R_I/R_F)| \ll 1,$$

by utilizing Equations 9.4 and 9.9, the *worst case limit of CMRR** can be given as

$$CMRR \geq -20\,\text{dB}\,\log_{10}\left[\,|A_{CM}/A| + \left|\left(1 - \frac{R_S}{R_P}\frac{R_F}{R_I}\right)\Big/\left(1 + \frac{R_F}{R_I}\right)\right|\,\right].$$

$$(9.14)$$

Example 9.3. An operational amplifier with a common mode rejection ratio of $CMRR = 80\,\text{dB}$ is used in the circuit of Fig. 9.2a. Nominally, $R_I = R_S = 1000\,\Omega$ and $R_F = R_P = 10{,}000\,\Omega$, but all four resistors have a $\pm 0.1\%$ tolerance. By utilizing Equation 9.5,

$$\left|\frac{A_{CM}}{A}\right| = 10^{-CMRR/20\,\text{dB}} = 10^{-80\,\text{dB}/20\,\text{dB}} = 10^{-4}.$$

The worst case limit of $CMRR$, from Equation 9.14, is given by

$$CMRR \geq -20\,\text{dB} \times \log_{10}\left[\,\left|\frac{A_{CM}}{A}\right| + \left|\left(1 - \frac{R_S}{R_P}\frac{R_F}{R_I}\right)\Big/\left(1 + \frac{R_F}{R_I}\right)\right|\,\right]$$

$$= -20\,\text{dB} \times \log_{10}\left[\,10^{-4} + \left|\left(1 - \frac{1001}{9990}\frac{10{,}010}{999}\right)\Big/\left(1 + \frac{10{,}010}{999}\right)\right|\,\right]$$

$$= -20\,\text{dB}\,\log_{10}\,(10^{-4} + 3.6 \times 10^{-4}) = 66.7\,\text{dB}.$$

INPUT IMPEDANCES

In the case of a real operational amplifier, input currents I_p and I_n of Fig. 2.1 are different from zero. Two approximately equivalent representations of the input terminals of a real operational amplifier

---

* It would seem that, based on Equation 9.9, a nonzero A_{CM} could be compensated by a suitable choice of parameters. Unfortunately, as a rule, only the magnitude of A_{CM} is known; hence, only the worst case limit of the common mode rejection ratio can be determined.

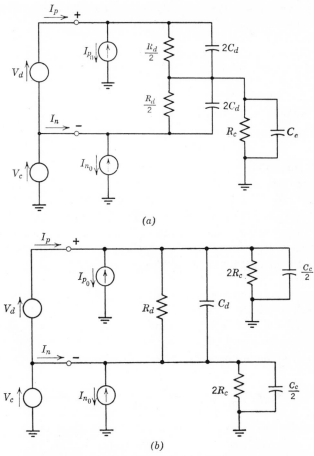

Figure 9.3. Two representations of the input terminals of an operational amplifier that are equivalent if $R_d \ll R_c$ and $C_d \gg C_c$.

are shown in Fig. 9.3. Input current I_p can be decomposed into a voltage-independent I_{p_0} and a voltage-dependent component. Input current I_n can be also decomposed into a voltage-independent I_{n_0} and a voltage-dependent component. In this section, the voltage-dependent components of the input currents will be discussed, while I_{p_0} and I_{n_0} will be discussed in the next chapter.

The *differential input impedance* of an operational amplifier represented by Fig. 9.3 can be defined as

$$Z_d \equiv \frac{\partial V_d}{\partial I_p}, \tag{9.15}$$

and the *common mode input impedance* as*

$$Z_c \equiv \frac{\partial V_c}{\partial (I_p + I_n)}. \tag{9.16}$$

Example 9.4.   At zero frequency, a Type 9406 operational amplifier has a differential input impedance of 7000 Ω and a common mode input impedance of 1 MΩ. Thus, in Fig. 9.3, $R_d = 7000$ Ω and $R_c = 1$ MΩ. It might be also useful to separate capacitances C_d and C_c of Fig. 9.3; unfortunately such a separation is rarely specified on present-day data sheets of operational amplifiers.

It is of significant interest to investigate the input impedance of feedback amplifier circuits. The noninverting feedback amplifier circuit of Fig. 3.1 with the input of the operational amplifier represented by Fig. 9.3b is shown in Fig. 9.4. It can be shown that the resulting input impedance at zero frequency, R_{in}, is

$$R_{\text{in}} \equiv \left(\frac{\partial V_{\text{in}}}{\partial I_p}\right)_{f=0} = \cfrac{1}{\cfrac{1}{R_c} + \cfrac{1}{\cfrac{R_I R_F}{R_I + R_F} + R_d \cfrac{A_{DC}}{M_{DC}}}}, \tag{9.17}$$

where M_{DC} is the resulting dc amplification of the feedback amplifier circuit.

Example 9.5.   A Type 9406 operational amplifier with $R_d = 7000$ Ω, $R_c = 1$ MΩ, and $A_{DC} = 1000$ is utilized as a noninverting feedback amplifier with $R_I = 1000$ Ω and

---

* When the component values in Fig 9.3 are constant, impedances Z_d and Z_c can be written as

$$Z_d = (V_d/I_p)_{V_c = I_{p_0} = I_{n_0} = 0} \quad \text{and} \quad Z_c = [V_c/(I_p + I_n)]_{V_d = I_{p_0} = I_{n_0} = 0}.$$

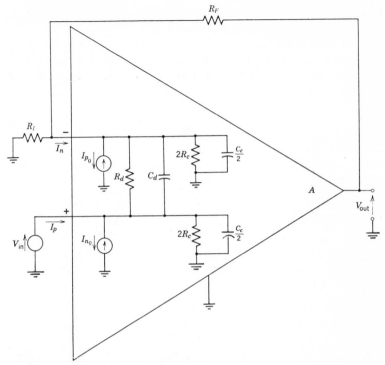

Figure 9.4. The noninverting feedback amplifier circuit of Fig. 3.1 with the input of the operational amplifier represented by the circuit of Fig. 9.3b.

$R_F = 9000\ \Omega$. Thus,

$$M_{DC} = \frac{A_{DC}}{1 + \dfrac{A_{DC}R_I}{R_I + R_F}} = \frac{1000}{1 + \dfrac{1000 \times 1000\ \Omega}{1000\ \Omega + 9000\ \Omega}} \approx 10,$$

and $A_{DC}/M_{DC} \approx 1000/10 \approx 100$. From Equation 9.17, the input impedance at zero frequency is

$$R_{in} \equiv \left(\frac{\partial V_{in}}{\partial I_p}\right)_{f=0}$$

$$= \left[\frac{1}{10^6\ \Omega} + \frac{1}{\dfrac{1000\ \Omega \times 9000\ \Omega}{1000\ \Omega + 9000\ \Omega} + 7000\ \Omega \times 100}\right]^{-1} = 0.41\ \text{M}\Omega.$$

OUTPUT IMPEDANCE

In the case of an ideal operational amplifier, Equation 2.1 states that the output voltage is determined by the input voltages, and is independent of the output current unless it is infinite. In reality, the output voltage is a function of the output current, that is, in Fig. 2.1,

$$Z_{out} \equiv -\frac{\partial V_{out}}{\partial I_{out}} \neq 0. \tag{9.18}$$

In general, *output impedance* Z_{out} is a function of frequency; in many cases it can be represented as a resistance in series with an inductance.

> **Example 9.6.** The output impedance of a Type 741 operational amplifier can be approximated by a resistance of 75 Ω in series with an inductance of 40 μH. Thus, $Z_{out}(f) = 75\ \Omega + \mathrm{j}2\pi\ 40\ \mu\mathrm{H} \times f$ and $|Z_{out}(f)| = \sqrt{75^2 + (2 \times 40 \times 10^{-6}f)^2}$. For example, at a frequency of $f = 1$ MHz,
>
> $$|Z_{out}(f)| = \sqrt{75^2 + (2 \times 40 \times 10^{-6} \times 10^6)^2} = 260\ \Omega.$$

When an operational amplifier with a finite output impedance is used in a feedback amplifier circuit (Fig. 9.5), the resulting output impedance of the circuit will be a function of the output impedance Z_{out} of the operational amplifier, of amplification A of the operational amplifier, and of resistors R_I and R_F. It can be shown that the resulting output impedance of the feedback amplifier circuit of Fig. 9.5 is

$$\frac{\partial V_{out}}{\partial I_{load}} = \frac{Z_{out}}{1 + \left(A + \dfrac{Z_{out}}{R_I}\right)\dfrac{R_I}{R_I + R_F}}. \tag{9.19}$$

If, as is the usual case,

$$\left|\frac{Z_{out}}{R_I}\right| \ll |A|, \tag{9.20}$$

then Equation 9.19 becomes

$$\frac{\partial V_{out}}{\partial I_{load}} \approx \frac{Z_{out}}{1 + A\dfrac{R_I}{R_I + R_F}} = \frac{Z_{out}}{1 + AF_N} = Z_{out}\frac{M_N}{A}. \tag{9.21}$$

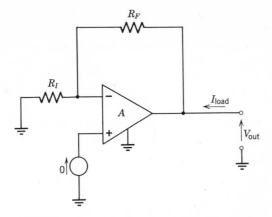

Figure 9.5. Circuit for determining the output impedance $\partial V_{out}/\partial I_{load}$ of a feedback amplifier circuit.

Example 9.7. A Type 741 operational amplifier with an output impedance of $Z_{out}(f) = 75\ \Omega + \text{j}2\pi 40\ \mu\text{H} \times f$ and an amplification of $A \approx 200{,}000/(1 + \text{j}f/10\ \text{Hz})$ is used in the circuit of Fig. 9.5 with $R_I = 100\ \Omega$ and $R_F = 10{,}000\ \Omega$. Hence, the feedback return is

$$F_N = \frac{R_I}{R_I + R_F} = \frac{100\ \Omega}{100\ \Omega + 10{,}000\ \Omega} \approx 0.01,$$

and the resulting amplification at zero frequency is

$$M_{DC} = \frac{A_{DC}}{1 + A_{DC}F_N} = \frac{200{,}000}{1 + 200{,}000 \times 0.01} \approx 100.$$

The resulting output impedance at zero frequency, by utilizing Equation 9.21, is thus

$$\left(\frac{\partial V_{out}}{\partial I_{load}}\right)_{f=0} = \left(Z_{out}\frac{M_N}{A}\right)_{f=0} = 75\ \Omega\,\frac{100}{200{,}000} = 0.0375\ \Omega.$$

SUPPLY VOLTAGE REJECTION

The output voltage of an ideal operational amplifier depends only on the input voltages and it is independent of the power supply voltage. In reality, the output voltage is a function of the power supply voltage, or voltages if there are more than one. It has been customary to define a *supply voltage rejection ratio*, or *Power Supply Rejection Ratio*, *PSRR*, as

$$PSRR \equiv \left| \frac{1}{A_{DC}} \frac{\partial V_{\text{out}}}{\partial V_{\text{supply}}} \right|, \qquad (9.22)$$

that is, as the ratio of the equivalent voltage change at the input of the amplifier to a change in supply voltage.

> **Example 9.8.** The supply voltage rejection ratio of a Type 741 operational amplifier is 30 μV/V = 30 × 10^{-6}, and its dc amplification is $A_{DC} = 200,000$. Thus, if there is a $\Delta V_{\text{supply}} = 10$ mV ripple on the power supply, this will be equivalent to a 10 mV × 30 × 10^{-6} = 0.3 μV on the input of the amplifier. Thus, without feedback, the ripple voltage on the output $|\Delta V_{\text{out}}| = 0.3$ μV × 200,000 = 60 mV. This result can be also obtained directly by utilizing Equation 9.22:

$$|\Delta V_{\text{out}}| = PSRR \times |A_{DC}| \times |\Delta V_{\text{supply}}|$$
$$= 30 \times 10^{-6} \times 200,000 \times 10 \text{ mV} = 60 \text{ mV}.$$

In the case of an operational amplifier with negative feedback and a resulting amplification of M_N, it can be shown that the output voltage ΔV_{out} resulting from a supply voltage change of ΔV_{supply} is given by

$$|\Delta V_{\text{out}}| = |PSRR \times M_N \times \Delta V_{\text{supply}}|. \qquad (9.23)$$

Example 9.9. A Type 741 operational amplifier has a supply voltage rejection ratio of $PSRR = 30$ μV/V and a dc amplification of $A_{DC} = 200,000$. It is utilized as a feedback amplifier with a resulting feedback amplification of $M_{DC} = 100$, and there is a $\Delta V_{\text{supply}} = 10$ mV ripple on the power supply voltage. As a result, by using Equation 9.23, the ripple at the output of the amplifier is

$$|\Delta V_{\text{out}}| = |PSRR \times M_N \times \Delta V_{\text{supply}}|$$
$$= 30 \times 10^{-6} \times 100 \times 10 \text{ mV}$$
$$= 30 \ \mu\text{V}.$$

In general, the supply voltage rejection ratio, *PSRR*, is a function of frequency having, as a rule, its best (lowest) value at dc; this frequency dependence is also a function of the compensation scheme used. Unfortunately, information on present-day operational amplifier data sheets on the frequency dependence of the supply voltage rejection ratio is very sketchy.

PROBLEMS

1. Common mode rejection properties of an operational amplifier are measured in the circuit of Fig. 9.1. At $V_c = 0$ and $V_d = 1$ mV, a $V_{out} = 5$ V is measured. At $V_c = 1$ mV and $V_d = 0$, a $V_{out} = -0.5$ mV is measured. Determine the values of A, $|A_{CM}|$, and the dc common mode rejection ratio *CMRR*.

2. A Type 741 operational amplifier has a dc common mode rejection ratio of $CMRR(f = 0) = 90$ dB. At a frequency of $f = 1$ MHz, $CMRR$ $(f = 1$ MHz$) = 16$ dB. Determine the value of corner frequency f_{CM} of the common mode rejection ratio.

3. Show that the common mode rejection ratio does not change if in the circuit of Fig. 9.1 the positive and negative input terminals of the operational amplifier are interchanged.

4. Show that the common mode rejection ratios of the circuits of Fig. 9.2a and Fig. 9.2b are identical if the components are identical.

5. An operational amplifier with a common mode rejection ratio of $CMRR = 100$ dB is used in the circuit of Fig. 9.2a. The useful signal is $V_d = 10 \, \mu V$ and the undesired noise is $V_c = 10$ mV. Determine the ratio of the useful signal to the undesired noise at the output of the circuit.

6. Derive Equations 9.9, 9.10, 9.11, 9.12, 9.13, and 9.14.

7. Show that the worst case limit of the common mode rejection ratio of the compound differential amplifier circuit of Fig. 3.8 can be approximated as

$$CMRR \approx -20 \text{ dB } \log_{10} \left[\left| \frac{A_{CM}}{A} \right| + \left| \left(1 - \frac{R_S}{R_P} \frac{R_F}{R_I} \right) \middle/ \left(1 + \frac{R_F}{R_I} \right) \right| \right.$$
$$\left. + |(1 - A_1/A_2)/A_1| \right],$$

provided that $A_1 \gg 1$, $A_2 \gg 1$, and

$$|(R_S/R_P - R_I/R_F)/(R_S/R_P + R_I/R_F)| \ll 1.$$

8. Derive Equation 9.17.

9. Estimate the input impedance seen at the positive input terminal of the circuit of Fig. 3.1 with the voltage source removed, if the operational amplifier input terminals can be represented by the circuit of Fig. 9.3 with $C_d = 10 \text{ pF}$, $C_c = 1 \text{ pF}$, $R_d = 10,000 \text{ }\Omega$, and $R_c = 1 \text{ M}\Omega$.

10. A Type 741 operational amplifier with an output impedance consisting of a resistance of 75 Ω in series with an inductance of 40 μH has an amplification of $A \approx 200,000/(1 + jf/10 \text{ Hz})$. Determine the resulting output impedance at zero frequency and the magnitude of the resulting output impedance at a frequency of $f = 1$ MHz if the amplifier is used as a voltage follower.

11. A Type 702A operational amplifier has a dc amplification of $A_{DC} = 4000$ and a supply voltage rejection ratio of $PSRR = 75 \text{ }\mu\text{V/V}$. Determine the ripple on the output, if the amplifier is used without feedback and if there is a ripple of 10 mV on the power supply voltage.

12. Derive Equation 9.23.

10

OOOOOOOOOOOOOOOOOOOOOOOOOOOOOOOO

Additional Properties of Real Operational Amplifiers

In the preceding chapter, linear attributes of real operational amplifiers were described. This chapter presents additional properties and limitations.

INPUT CURRENTS

The input circuit of an operational amplifier has been represented by the circuit of Fig. 9.3. It can be seen that when $V_c = V_d = 0$, that is, when the voltages on both input terminals of the operational amplifier are zero, there is a current I_{p_0} flowing into the positive terminal and a current I_{n_0} into the negative terminal. For many practical operational amplifiers,

$$|I_{p_0} - I_{n_0}| \ll \left| \frac{I_{p_0} + I_{n_0}}{2} \right|. \tag{10.1}$$

Thus, it is reasonable to define an *input bias current* I_B as the average of I_{p_0} and I_{n_0}:

$$I_B \equiv \frac{I_{p_0} + I_{n_0}}{2}, \tag{10.2}$$

and an *input offset current* I_{OFF} as

$$I_{\mathrm{OFF}} \equiv I_{p_0} - I_{n_0}.^* \tag{10.3}$$

Example 10.1. At a temperature of 25°C, a Type 741 operational amplifier has a typical input bias current of $I_{B_{\mathrm{typ}}} = 80$ nA, a maximum input bias current of $I_{B_{\mathrm{max}}} = 500$ nA, a typical input offset current of $|I_{\mathrm{OFF}}|_{\mathrm{typ}} = 20$ nA, and a maximum input offset current of $|I_{\mathrm{OFF}}|_{\mathrm{max}} = 200$ nA. No minimum is specified for I_B, and it is reasonable to assume that this minimum is zero. Thus, if the current into the positive input terminal is $I_p = 250$ nA, then the current into the negative input terminal, I_n, can be between 50 nA and 450 nA; if $I_p = 600$ nA, then $I_n = 400$ nA; and if $I_p = 10$ nA, then I_n can be between zero and 210 nA.

When operation over a certain temperature range is desired, it is important to take into account the temperature dependence of the input currents.

Example 10.2. At a temperature of -55°C, a Type 741 operational amplifier has a maximum input bias current of $I_{B_{\mathrm{max}}} = 1.5$ μA and a maximum input offset current of $|I_{\mathrm{OFF}}|_{\mathrm{max}} = 0.5$ μA. At a temperature of $+125$°C, $I_{B\mathrm{max}} = 0.5$ μA and $|I_{\mathrm{OFF}}|_{\mathrm{max}} = 0.2$ μA. Comparing with the data in Example 10.1 above, $I_{B_{\mathrm{max}}}$ and $|I_{\mathrm{OFF}}|_{\mathrm{max}}$ are the same at $+125$°C as they are at $+25$°C, but are worse at -55°C.

In some cases the temperature dependence is given in terms of a *temperature coefficient*. If the temperature coefficient η of a current is measured in amperes per degree centigrade (°C), then the current change ΔI over a temperature range ΔT can be approximated as

$$\Delta I = \eta \, \Delta T. \tag{10.4}$$

---

* Frequently, the input offset current is specified as $|I_{\mathrm{OFF}}|$ or as $\pm|I_{\mathrm{OFF}}|$. Also, in some cases input currents I_{p_0} and I_{n_0} of Equation 10.2 are specified separately.

Example 10.3. The maximum temperature coefficient of the input offset current of an operational amplifier is Max $|\eta_{\text{OFF}}| = 0.1$ nA/°C. Thus, if the temperature varies by $\Delta T = 10$°C, the input offset current will vary by $\Delta I = |\eta_{\text{OFF}}| \Delta T = 0.1$ nA/°C $\times$ 10°C $= 1$ nA, or by less.

When an operational amplifier with an input bias current of I_B and an input offset current of I_{OFF} is used in the feedback amplifier circuit of Fig. 3.4, it can be shown that, for $V_p = V_n = 0$, V_{out} can be approximated as

$$V_{\text{out}} \approx -M_{DC} \frac{R_I R_F}{R_I + R_F} I_{\text{OFF}}$$

$$+ M_{DC}\left(\frac{R_I R_F}{R_I + R_F} - \frac{R_S R_P}{R_S + R_P}\right)I_B, \quad (10.5a)$$

where the resulting dc amplification of the feedback amplifier circuit, M_{DC}, is defined as

$$M_{DC} \equiv \frac{A_{DC}}{1 + \dfrac{A_{DC}R_I}{R_I + R_F}}, \quad (10.5b)$$

and A_{DC} is the amplification of the operational amplifier at zero frequency.

Example 10.4. At a temperature of 25°C, a Type 2741 operational amplifier has a maximum input offset current of $|I_{\text{OFF}}| = 50$ pA, a maximum input bias current of $|I_B| = 100$ pA, and a dc amplification of $A_{DC} = 30,000$. The amplifier is used in the circuit of Fig. 3.4 with $R_S = R_I = 10$ MΩ, $R_P = R_F = 90$ MΩ, and $V_p = V_n = 0$. Thus, the resulting dc amplification of the feedback amplifier circuit at zero frequency, M_{DC}, is

$$M_{DC} = \frac{A_{DC}}{1 + \dfrac{A_{DC}R_I}{R_I + R_F}} = \frac{30,000}{1 + \dfrac{30,000 \times 10 \text{ MΩ}}{10 \text{ MΩ} + 90 \text{ MΩ}}} \approx 10.$$

Since $R_S = R_I$ and $R_P = R_F$, input bias current I_B has no effect on the output voltage (see Equation 10.5a). Input offset current I_{OFF} will result in an output voltage of

$$|V_{\text{out}}| \leq \left| -M_{DC} \frac{R_I R_F}{R_I + R_F} I_{OFF} \right|$$

$$= \left| -10 \frac{10 \text{ M}\Omega \times 90 \text{ M}\Omega}{10 \text{ M}\Omega + 90 \text{ M}\Omega} 50 \text{ pA} \right| = 4.5 \text{ mV}.$$

INPUT OFFSET VOLTAGE

In the case of an ideal operational amplifier, the output voltage will be zero if both input voltages are zero. In the case of a real operational amplifier, however, there may be a nonzero output voltage even if both input voltages are zero. It has been customary to define, for the operational amplifier of Fig. 2.1 with a dc amplification of A_{DC}, an *input offset voltage* V_{OFF} as

$$V_{\text{OFF}} = \left| \frac{V_{\text{out}}(V_p = 0, V_n = 0)}{A_{DC}} \right|. \tag{10.6}$$

Thus, V_{OFF} is an equivalent offset voltage at the input of the amplifier.* The input offset voltage can be also represented by a battery V_{OFF} connected in series with one of the input terminals of the operational amplifier.

> **Example 10.5.** The input offset voltage of an operational amplifier with a dc amplification of $A_{DC} = 10,000$ is measured by grounding both input terminals and measuring the output voltage. An output voltage of $V_{\text{out}} = 5 \text{ V}$ is measured this way. Thus, the input offset voltage is $V_{\text{OFF}} = 5 \text{ V}/10,000 = 0.5 \text{ mV}.$

The input offset voltage V_{OFF} is, in general, a function of temperature. This temperature dependence can be described either by specifying V_{OFF} at several temperatures, or by a *temperature coefficient* $|dV_{OFF}/dT|$, where T is the temperature.

---

*Frequently, the input offset voltage is specified as $\pm V_{\text{OFF}}$.

Example 10.6. The temperature coefficient of the input offset voltage V_{OFF} of a Type 702A operational amplifier is specified as being less than 10 μV/°C between the temperatures of -55°C and $+125$°C. Thus, if the temperature changes from 0°C to $+50$°C, input offset voltage V_{OFF} will change by 50°C $\times$ 10 μV/°C = 500 μV, or by less. If the operational amplifier is operated without feedback, then, since its dc amplification is $A_{DC} = 4000$, the output voltage will change by 500 μV $\times$ 4000 = 2 V, or by less, over the 0°C to 50°C temperature range.

If the operational amplifier is used as a feedback amplifier in the circuit of Fig. 3.4 with $V_p = V_n = 0$, then it can be shown that an input offset voltage V_{OFF} results in an output voltage of

$$V_{\text{out}} = M_{DC}V_{OFF}, \tag{10.7a}$$

where the resulting dc amplification, M_{DC}, is defined as

$$M_{DC} \equiv \frac{A_{DC}}{1 + A_{DC}\dfrac{R_I}{R_I + R_F}}, \tag{10.7b}$$

and A_{DC} is the amplification of the operational amplifier at zero frequency.

LIMITATIONS AND RATINGS

In the preceding, it was assumed that the dc amplification A_{DC} of the operational amplifier was constant. In reality, ignoring the offset voltage, the output voltage as function of input voltage can be characterized by a curve such as the one shown in Fig. 10.1. It can be seen that the slope of the curve, that is, dc amplification $A_{DC} \equiv \partial V_{\text{out}}/\partial V_{\text{in}}$, is a fairly constant $A_{DC} = 10,000$ between a *maximum output voltage* of approximately $+20$ V and a *minimum output voltage* of approximately -10 V. A realistic design must take these limiting voltages ("*output voltage swing*") into account.

Additional limitations are imposed by the maximum and minimum values of the common mode input voltage (V_c in Fig. 9.2) the amplifier is capable of tolerating.

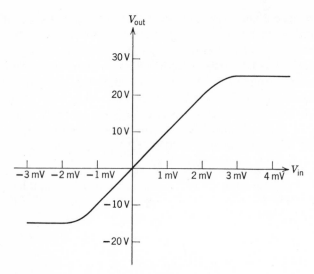

Figure 10.1. Output voltage V_{out} versus input voltage V_{in} for a real operational amplifier.

Example 10.7. A Type 702A operational amplifier has a maximum output voltage swing of ± 5 V, a minimum input voltage of -6 V, and a maximum input voltage of $+1.5$ V. The amplifier is used in the circuit of Fig. 9.2a with $R_S = R_I = 1000\ \Omega$ and $R_P = R_F = 9000\ \Omega$. The value of the dc amplification of the operational amplifier is $A_{DC} = 4000$, and it can be shown that the resulting dc amplification of the feedback amplifier circuit is $M_{DC} \approx 10$.

The limitation of ± 5 V on the output results in a limitation of $\pm 5\ \text{V}/M_{DC} = \pm 5\ \text{V}/10 = \pm 0.5$ V on the differential input voltage. Observing the minimum input voltage limitation of -6 V, if the full output voltage swing is to be utilized, the common mode input voltage V_c must be more positive than -5.5 V; also, as a result of the maximum input voltage limitation of $+1.5$ V, common mode input voltage V_c must be more negative than $+1$ V.

Particular care should be exercised in the design to assure that the *maximum ratings* of an operational amplifier are not exceeded. Such ratings include, although are not restricted to, maximum supply voltages, peak output current, maximum and minimum input voltages, maximum differential input voltage, power dissipation, operating temperature range, storage temperature range, and lead temperature during soldering. In general, damage to the structure of the operational amplifier may result if a maximum rating is exceeded.

SLEW RATE

The slew rate is a limitation on the rate of change in the output voltage of an operational amplifier. The source of this limitation can be seen in Example 5.1, where a stage of an operational amplifier is represented by the circuit of Fig. 5.1 consisting of a 1-mA current source parallel with a resistance of $R = 1000 \ \Omega$ and with a capacitance C. It is seen that for a step-function input current of $I_{in} = 1 \text{ mA} \times u(t)$, the voltage is $V_{out} = 1 \text{ V}(1 - e^{-t/RC})$. In a real operational amplifier, the source of current I_{in} is a transistor that can not deliver arbitrarily large currents. Thus, I_{in}, and hence dV_{out}/dt, are limited. Such limitations lead to a specification of the output slew rate, usually given in units of V/μs.

Example 10.8. In the circuit of Fig. 5.1, capacitance $C = 10$ pF, and the maximum available input current is $I_{in} = 1$ mA. Voltage V_{out} can be written as

$$V_{out} = I_{in}R(1 - e^{-t/RC}).$$

From this,

$$\frac{dV_{out}}{dt} = \frac{I_{in}}{C} e^{-t/RC}.$$

The slew rate S is the maximum of $|dV_{out}/dt|$,

$$S = \left| \frac{dV_{out}}{dt} \right|_{max} = \left| \frac{I_{in}}{C} \right|_{max} = \left| \frac{1 \text{ mA}}{10 \text{ pF}} \right| = 100 \text{ V}/\mu s.$$

The finite slew rate also imposes a limitation on the maximum amplitude of a sinewave the operational amplifier can deliver at its output. In the case of a sinewave in the form

$$V_s = V_0 \sin 2\pi ft, \tag{10.8}$$

the rate of change in voltage V_s is

$$\frac{dV_s}{dt} = V_0 2\pi f \cos 2\pi f t. \tag{10.9}$$

The maximum value of $|dV_s/dt|$ is limited by the slew rate S:

$$S = \left| \frac{dV_s}{dt} \right|_{\max} \geqq V_0 2\pi f; \tag{10.10}$$

hence, the maximum of amplitude V_0, $V_{0_{\max}}$, that is available at a frequency f is given by

$$V_{0_{\max}} = \frac{S}{2\pi f}. \tag{10.11}$$

Example 10.9. A Type 107 internally compensated operational amplifier has a slew rate of 0.5 V/μs. Thus, for a sinewave with a frequency of $f = 10$ kHz, the maximum amplitude available at the output of the amplifier is

$$V_{0_{\max}} = \frac{S}{2\pi f} = \frac{0.5 \text{ V}/\mu\text{s}}{2\pi 10 \text{ kHz}} = 8 \text{ V}.$$

In general, the slew rate and the maximum output amplitude as function of frequency depend on the feedback amplification and on the type of frequency compensation used. For this reason, the slew rate is usually given for a specific circuit or circuits.

NOISE

Electrical conduction takes place by means of discrete charge carriers, such as electrons and holes. As a result, noise voltage and noise currents are superimposed on the inputs of the operational amplifier. When the signal levels are low, these noise sources may become significant.

If the operational amplifier "sees" a resistance R at its input terminals, the resulting total input noise power per unit bandwidth, designated as *narrow-band* or *spot noise*, is given by

$$\frac{v_t^2}{R} = 4kT + \frac{v_n^2}{R} + R i_n^2. \tag{10.12}$$

The term $4kT$ is the thermal noise power per unit bandwidth contributed by resistance R, v_n^2/R is the noise power per unit bandwidth contributed by the *input noise voltage* v_n of the operational amplifier, and Ri_n^2 is the noise power per unit bandwidth contributed by the *input noise current* i_n of the operational amplifier. The value of kT at room temperature is 0.4×10^{-20} VA/Hz; v_t and v_n are measured in $V/\sqrt{Hz}$, i_n in $A/\sqrt{Hz}$.

When v_n and i_n can be considered constants within the bandwidth of interest B, the resulting *input noise power*, P_B, is given by

$$P_B = Bv_t^2/R, \tag{10.13}$$

where v_t^2/R is given by Equation 10.12.* Also, the *rms input noise voltage*, V_B, can be written as

$$V_B = \sqrt{P_B R} = v_t\sqrt{B}. \tag{10.14}$$

Example 10.10. At a frequency of 10 kHz, an operational amplifier has an input noise voltage of $v_n = 10$ nV/$\sqrt{Hz}$ and an input noise current of $i_n = 1$ pA/$\sqrt{Hz}$, both constant within the bandwidth of interest, $B = 100$ Hz. It has an amplification of $A = 10,000$ and it is operated in the inverting amplifier circuit of Fig. 3.2 with $R_I = 1.01$ kΩ and $R_F = 100$ kΩ. Thus, the resulting amplification is $M_I = -100$ and the resistance seen by the operational amplifier at its input terminals is $R = 1$ kΩ.

The resulting total input noise power per unit bandwidth is given by Equation 10.12 as

$$\frac{v_t^2}{R} = 4kT + \frac{v_n^2}{R} + Ri_n^2$$

$$= 1.6 \times 10^{-20} \text{ VA/Hz} + \frac{10^{-16} \text{ V}^2/\text{Hz}}{1000 \ \Omega}$$

$$+ 1000 \ \Omega \ 10^{-24} \text{ A}^2/\text{Hz}$$

$$= 1.161 \times 10^{-19} \text{ VA/Hz}.$$

---

* When v_n or i_n can not be considered constant, the resulting *wideband noise* has to be determined by integrating over the bandwidth the product of the amplification and the input noise power per unit bandwidth.

The resulting input noise power within the bandwidth of $B = 100$ Hz, from Equation 10.13, is

$$P_B = B v_t^2 / R = 100 \text{ Hz } 1.161 \times 10^{-19} \text{ VA/Hz}$$
$$= 1.161 \times 10^{-17} \text{ VA},$$

and the resulting rms noise voltage at the input, from Equation 10.14, is

$$V_B = \sqrt{P_B R} = \sqrt{1.161 \times 10^{-17} \text{ VA } 1000 \,\Omega} = 108 \text{ nV}.$$

The resulting rms noise voltage at the output of the amplifier is $|M_I| \, V_B = 100 \times 108 \text{ nV} = 10.8 \,\mu\text{V}$.

In many cases, the noise performance of an amplifier circuit is described by a *noise figure* $\mathscr{F}$, which is a measure of the noise degradation resulting from adding the noise of the operational amplifier to the thermal noise of input resistance R. When the resulting amplification is large ($|M_I| \gg 1$ or $M_N \gg 1$), the noise figure can be approximated as

$$\mathscr{F} = 10 \text{ dB } \log_{10} \left(1 + \frac{v_n^2 / R + i_n^2 R}{4kT} \right). \qquad (10.15)$$

It can be shown that the noise figure $\mathscr{F}$ has its minimum value, $\mathscr{F}_{\min}$, when $R = R_{\text{opt}} = v_n / i_n$, and it is

$$\mathscr{F}_{\min} = \mathscr{F}_{R = v_n / i_n} = 10 \text{ dB } \log_{10} \left(1 + \frac{v_n^2 / R_{\text{opt}}}{2kT} \right). \qquad (10.16)$$

Example 10.11. In the preceding example, $v_n = 10^{-8}$ V/$\sqrt{\text{Hz}}$, $i_n = 10^{-12}$ A/$\sqrt{\text{Hz}}$, $R = 1000 \,\Omega$, and $|M_I| \gg 1$. Thus, the noise figure, from Equation 10.15, and with $kT = 0.4 \times 10^{-20}$ VA/Hz,

$$\mathscr{F} = 10 \text{ dB } \log_{10} \left(1 + \frac{v_n^2 / R + i_n^2 R}{4kT} \right)$$

$$= 10 \text{ dB } \log_{10} 7.4 = 8.7 \text{ dB}.$$

The minimum noise figure is attained at $R = R_{\text{opt}} = v_n / i_n = (10^{-8} \text{ V} / \sqrt{\text{Hz}}) / (10^{-12} \text{ A} / \sqrt{\text{Hz}}) = 10 \text{ k}\Omega$, and its value, from Equation 10.16, is

$$\mathscr{F}_{\min} = 10 \text{ dB } \log_{10} \left(1 + \frac{v_n^2/R_{\text{opt}}}{2\text{k}T} \right)$$

$$= 10 \text{ dB } \log_{10} \left(1 + \frac{(10^{-16} \text{ V}^2/\text{Hz})/10{,}000 \text{ } \Omega}{2 \times 0.4 \times 10^{-20} \text{ VA/Hz}} \right)$$

$$= 10 \text{ dB } \log_{10} 2.25 = 3.5 \text{ dB}.$$

Above considerations apply equally well to inverting and non-inverting amplifier circuits. In the case of differential amplifier circuits, however, the evaluation of the noise is slightly more involved.*

PROBLEMS

1. At a temperature of $+25°C$, a Type 9406 operational amplifier has a maximum input bias current of ± 10 μA and a maximum input offset current of ± 1 μA. Determine the maximum and minimum values of the current I_p into the positive input terminal, if $I_n = 0.5$ μA, 5 μA, and 10.5 μA.

2. The Type 2741 operational amplifier uses field-effect transistors at its inputs and as a result it has low input currents. At a temperature of $+25°C$, the maximum input bias current is 100 pA and the maximum input offset current is 50 pA. The signs of the currents are not specified and it will be assumed that they can be either positive or negative. Three of these amplifiers are utilized in the circuit of Fig. 3.8 with $R_1 = 10$ MΩ, $R_2 = 11$ MΩ, $R_I = R_S = 100$ Ω, $R_F = R_P = 10{,}000$ Ω, and $A_1 = A_2 = A_3 = 30{,}000$. Determine the maximum change in output voltage V_{out} resulting from the input bias currents and from the input offset currents.

3. Derive Equation 10.5.

4. The input offset current of a Type 9406 operational amplifier has a maximum temperature coefficient of 0.1 $\mu A/°C$. Determine the minimum and maximum values of the input offset current I_{OFF}, if

---

* For more details, and for a computation of wideband noise, see A. C. Markkula, Jr., Practical Considerations in the Design of Systems Using Linear Integrated Circuits, in *Application Considerations for Linear Integrated Circuits*, Edited by J. Eimbinder, Wiley-Interscience, New York, 1970.

the temperature is varied between $-55°C$ and $+125°C$, and if at a temperature of $+25°C$ the input offset current is $I_{OFF} = 1 \ \mu A$.

5. The input offset voltage V_{OFF} of a Type 2741 operational amplifier has a temperature coefficient $|dV_{OFF}/dT|$ of less than 25 $\mu V/°C$; the dc amplification is $A_{DC} = 30,000$. Determine the maximum change in the output voltage as a result of the finite dV_{OFF}/dT, if the amplifier is operated without feedback.

6. Derive Equation 10.7.

7. At a temperature of $+25°C$, the maximum input offset voltage of a Type 2741 operational amplifier is $V_{OFF} = 5$ mV. This operational amplifier is used in the differential amplifier circuit of Fig. 3.4 with $V_n = V_p = 0$, $R_S = R_I = 10 \ M\Omega$, and $R_P = R_F = 90 \ M\Omega$. Determine the maximum change in output voltage V_{out} as a result of input offset voltage V_{OFF}.

8. At a temperature of $+25°C$, a Type 107 operational amplifier has a maximum input bias current of 75 nA, a maximum input offset current of 10 nA, a maximum input offset voltage of 2 mV, and a dc amplification of $A_{DC} = 160,000$. Determine the range of the output voltage, if the amplifier is used in the circuit of Fig. 3.4 with $R_S = R_I = 10 \ M\Omega$, $R_P = R_F = 90 \ M\Omega$, and $V_p = V_n = 0$.

9. A Type 9406 internally compensated operational amplifier has a slew rate of $\pm 360 \ V/\mu s$. Determine the maximum frequency at which this amplifier can supply a 10-V *peak to peak* sinewave.

10. When lead-lag compensation is applied to a Type 702A operational amplifier, for an $M_{DC} = 10$ its slew rate is 50 V/μs. The maximum and minimum voltages the amplifier can provide at its output are $+5$ V and -5 V, respectively. Sketch the maximum available sinewave amplitude as function of frequency for frequencies between 1 kHz and 10 MHz.

11. Sketch the output voltage waveform of an amplifier with a dc amplification of $A_{DC} = 10,000$ and a slew rate of $S = 1 \ V/\mu s$, if it is used in the circuit of Fig. 2.2 with V_{in} a 1-MHz *square-wave* with a peak-to-peak amplitude of 1 V.

12. The noise characteristics of an operational amplifier are specified at a frequency of 1 kHz by $v_n = 1 \ nV/\sqrt{Hz}$ and $i_n = 1 \ pA/\sqrt{Hz}$. It has an amplification of $A = 10,000$ and it is used in the noninverting feedback amplifier circuit of Fig. 3.1 with a resulting amplification of $M_N = 100$ and with a bandwidth of $B = 10$ Hz. Find the values of input resistor R_I and feedback resistor R_F such that the minimum noise figure is attained. Find the value of the minimum noise figure. What is the resulting rms noise voltage at the output of the amplifier circuit?

13. Show that the noise figure of Equation 10.15 has its minimum value when $R = v_n/i_n$.

14. Show that the rms input noise voltage is smallest when the value of resistance R in Equation 10.12 is the smallest. Identify the reason why $R = 0$ is not feasible.

Appendix

Table 1. Amplification formulae for feedback amplifier circuits

Configuration	NONINVERTING	INVERTING
Schematic diagram		
Feedback return	$F_N \equiv \dfrac{R_I}{R_I + R_F}$	$F_I \equiv \dfrac{R_I}{R_F}$
Feedback factor	AF_N	AF_I
Definition of amplification	$M_N \equiv \dfrac{V_{\text{out}}}{V_{\text{in}}}$	$M_I \equiv \dfrac{V_{\text{out}}}{V_{\text{in}}}$
Amplification	$M_N = \dfrac{A}{1 + AF_N}$	$M_I = \dfrac{-A}{1 + (A + 1)F_I}$

	$M_N \approx 1/F_N$	$M_I \approx -1/F_I$
Amplification for large feedback factor		
Sensitivity to changes in A	$\dfrac{\Delta M_N}{M_{N\text{nom}}} = \dfrac{1}{1 + A_{\text{nom}}F_N}\dfrac{\Delta A}{A_{\text{nom}}}$	$\dfrac{\Delta M_I}{M_{I\text{nom}}} = \dfrac{1 + F_I}{1 + (A_{\text{nom}} + 1)F_I}\dfrac{\Delta A}{A_{\text{nom}}}$
Sensitivity to changes in A for large feedback factor	$\dfrac{\Delta M_N}{M_{N\text{nom}}} \approx \dfrac{1}{A_{\text{nom}}F_N}\dfrac{\Delta A}{A_{\text{nom}}}$	$\dfrac{\Delta M_I}{M_{I\text{nom}}} \approx \dfrac{1 + F_I}{A_{\text{nom}}F_I}\dfrac{\Delta A}{A_{\text{nom}}}$
Sensitivity to changes in feedback resistor R_F	$\dfrac{\Delta M_N}{M_{N\text{nom}}} = M_{N\text{nom}}F_{N\text{nom}}(1 - F_{N\text{nom}})\dfrac{\Delta R_F}{R_{F\text{nom}}}$	$\dfrac{\Delta M_I}{M_{I\text{nom}}} = -M_{I\text{nom}}F_{I\text{nom}}\dfrac{\Delta R_F}{R_{F\text{nom}}}$
Sensitivity to changes in feedback resistor R_F for large feedback factor	$\dfrac{\Delta M_N}{M_{N\text{nom}}} \approx (1 - F_{N\text{nom}})\dfrac{\Delta R_F}{R_{F\text{nom}}}$	$\dfrac{\Delta M_I}{M_{I\text{nom}}} \approx \dfrac{\Delta R_F}{R_{F\text{nom}}}$
Sensitivity to changes in input resistor R_I	$\dfrac{\Delta M_N}{M_{N\text{nom}}} = -M_{N\text{nom}}F_{N\text{nom}}(1 - F_{N\text{nom}})\dfrac{\Delta R_I}{R_{I\text{nom}}}$	$\dfrac{\Delta M_I}{M_{I\text{nom}}} = M_{I\text{nom}}F_{I\text{nom}}\dfrac{\Delta R_I}{R_{I\text{nom}}}$
Sensitivity to changes in input resistor R_I for large feedback factor	$\dfrac{\Delta M_N}{M_{N\text{nom}}} \approx -(1 - F_{N\text{nom}})\dfrac{\Delta R_I}{R_{I\text{nom}}}$	$\dfrac{\Delta M_I}{M_{I\text{nom}}} \approx -\dfrac{\Delta R_I}{R_{I\text{nom}}}$

Table 2. Network Transfer Functions

Network	Schematic Diagram	Transfer Function
Resistive divider		$$\frac{V_{\text{out}}(f)}{V_{\text{in}}(f)} = \frac{R_P}{R_S + R_P}$$
Lag network		$$\frac{V_{\text{out}}(f)}{V_{\text{in}}(f)} = \frac{1}{1 + jf/f_0};$$ $$f_0 \equiv \frac{1}{2\pi RC}$$
Modified lag network		$$\frac{V_{\text{out}}(f)}{V_{\text{in}}(f)} = \frac{1 + jf/f_2}{1 + jf/f_1};$$ $$f_1 \equiv \frac{1}{2\pi(R_1 + R_2)C},$$ $$f_2 \equiv \frac{1}{2\pi R_2 C}$$

Lead network	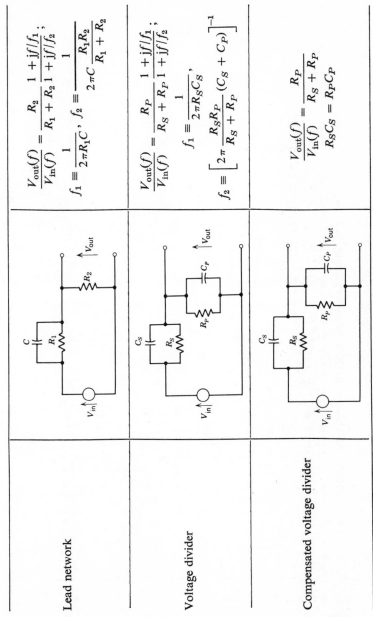	$\dfrac{V_{\text{out}}(f)}{V_{\text{in}}(f)} = \dfrac{R_2}{R_1+R_2}\dfrac{1+jf/f_1}{1+jf/f_2};$ $f_1 \equiv \dfrac{1}{2\pi R_1 C},\ f_2 \equiv \dfrac{1}{2\pi C\,\dfrac{R_1 R_2}{R_1+R_2}}$
Voltage divider		$\dfrac{V_{\text{out}}(f)}{V_{\text{in}}(f)} = \dfrac{R_P}{R_S+R_P}\dfrac{1+jf/f_1}{1+jf/f_2}$ $f_1 \equiv \dfrac{1}{2\pi R_S C_S},$ $f_2 \equiv \left[2\pi\,\dfrac{R_S R_P}{R_S+R_P}(C_S+C_P)\right]^{-1}$
Compensated voltage divider		$\dfrac{V_{\text{out}}(f)}{V_{\text{in}}(f)} = \dfrac{R_P}{R_S+R_P}$ $R_S C_S = R_P C_P$

Table 3. Stability Criteria of Amplifier Circuits with Negative Feedback

	AF_N (noninverting amplifier; Fig. 3.1) or $\dfrac{AF_I}{1+F_I}$ (inverting amplifier; Fig. 3.2)	CRITERION OF STABILITY $(K > 0)^*$
1.	$\dfrac{K}{1+jf/f_0}$	$K < \infty$
2.	$\dfrac{K}{(1+jf/f_1)(1+jf/f_2)}$	$K < \infty$
3.	$\dfrac{K}{(1+jf/f_0)^3}$	$K < 8$
4.	$\dfrac{K}{(1+jf/f_1)^2(1+jf/f_2)}$	$K < 4 + 2\left(\dfrac{f_2}{f_1}+\dfrac{f_1}{f_2}\right)$
5.	$\dfrac{K}{(1+jf/f_1)(1+jf/f_2)(1+jf/f_3)}$	$K < 2 + \dfrac{f_2+f_3}{f_1} + \dfrac{f_1+f_3}{f_2} + \dfrac{f_1+f_2}{f_3}$
6.	$\dfrac{K}{(1+jf/f_0)^4}$	$K < 4$

7.	$\dfrac{K}{(1 + jf/f_1)^3(1 + jf/f_2)}$	$K < 8\dfrac{(1 + f_1/f_2)^3}{(1 + 3f_1/f_2)^2}$
8.	$\dfrac{K}{jf/f_0(1 + jf/f_1)(1 + jf/f_2)^2}$	$K < \dfrac{2f_2}{f_0}\dfrac{(1 + f_2/f_1)^2}{(1 + 2f_2/f_1)^2}$
9.	$\dfrac{K}{jf/f_0(1 + jf/f_1)(1 + jf/f_2)(1 + jf/f_3)}$	$K < \dfrac{2f_1f_2f_3 + f_1^2f_2 + f_1^2f_3 + f_2^2f_1 + f_2^2f_3 + f_3^2f_1 + f_3^2f_2}{f_0(f_1 + f_2 + f_3)^2}$
10.	$\dfrac{K(1 + jf/f_1)^2}{(jf/f_0)^3}$	$K > \dfrac{1}{2}\left(\dfrac{f_1}{f_0}\right)^3$

* K is the dc feedback factor, a real positive number.

Table 4. Typical Properties of Operational Amplifiers Used in the Examples and Problems

Type	107	702A	741	2741	9406
Description	Internally compensated, Monolithic	Monolithic	Internally compensated, Monolithic	Field-effect transistor input, Hybrid	Internally compensated, Hybrid
DC Amplification	160,000	4000	200,000	30,000	1000
Maximum input bias current	75 nA	5 μA	0.5 μA	100 pA	10 μA
Maximum input offset current	10 nA	0.5 μA	0.2 μA	50 pA	1 μA
Maximum input offset voltage	2 mV	2 mV	5 mV	5 mV	10 mV
Maximum temperature coefficient of input offset voltage	15 μV/°C	10 μV/°C	30 μV/°C	25 μV/°C	100 μV/°C
DC common mode rejection ratio	96 dB	100 dB	90 dB	70 dB	80 dB
Differential input resistance	4 MΩ	40 kΩ	2 MΩ	100 GΩ	7 kΩ
Supply voltage rejection ratio	15 μV/V	75 μV/V	30 μV/V	100 μV/V	3 mV/V
Corner frequencies	5 Hz 10 MHz	1 MHz 4 MHz 40 MHz	10 Hz 10 MHz	10 Hz 1 MHz	1.5 MHz 150 MHz
Slew rate	0.5 V/μs	50 V/μs	0.5 V/μs	5 V/μs	360 V/μs

Answers to Selected Problems

OOOOOOOOOOOOOOOOOOOOOOOOOOOOOOOOOOO

Chapter 1
1. −1 V
3. −45°

Chapter 2
7. 0, 10 V
11. 1 second

Chapter 3
10. 0, 10 mV
14. −0.999 mV
16. −10,000

Chapter 4
10. 0
12. 10%

Chapter 5
12. 3.18 MHz

Chapter 8
2. 0.18 μF
4. 2 inches ($\approx$ 5 cm)

Chapter 9
1. 5000, 0.5, 80 dB
2. 200 Hz

Chapter 10
7. 45 mV
8. ±830 mV

Index

○○○○○○○○○○○○○○○○○○○○○○○○○○○○○○○○○

145